# WATERSHEDS

## A PRACTICAL HANDBOOK FOR HEALTHY WATER

Illustrations and original concept by Clive Dobson

Text by Gregor Gilpin Beck

FIREFLY BOOKS

# A FIREFLY BOOK

Published by Firefly Books Ltd., 1999

**Second Printing 2003**

*Canadian Cataloguing in Publication Data*

Dobson, Clive, 1949–    .
    Watersheds: a practical handbook for healthy water

Includes bibliographical references and index.
ISBN 1–55209–330–1

1. Watersheds.    2. Valley ecology.    3. Watershed management.
4. Water conservation.    5. Ecology.
I. Beck, Gregor Gilpin. 1964–    II. Title.

TC409.D62 1999        333.73        C98–932661–6

Published in Canada in 1999 by
Firefly Books Ltd.
3680 Victoria Park Avenue
Toronto, Ontario
Canada M2H 3K1

Published in the United States in 1999 by
Firefly Books (U.S.) Inc.
P.O. Box 1338, Ellicott Station
Buffalo, New York
USA 14205

*Library of Congress Cataloging-in-Publication Data*

Dobson, Clive, 1949–    .
    Watersheds: a practical handbook for healthy water /
Clive Dobson, Gregor Gilpin Beck.

    [152] p.: col. ill., maps;    cm.
Includes bibliographical references (p.    ) and index.
Summary: Introduction to environmental issues viewed through the ecology of the watershed.

ISBN 1–55209–330–1

1. Watersheds.    2. Ecology.    3. Environmental protection.
I. Beck, Gregor Gilpin, 1964–    .    II. Title.

333.73---dc21        1999        CIP

Produced by: Denise Schon Books Inc.
Design: Counterpunch/Linda Gustafson + Sue Becker
Editor: Lorraine Johnson

Printed and bound in Canada by Friesens, Altona, Manitoba

*The Publisher acknowledges the financial support of the Government of Canada through the Book Publishing Industry Development Program for its publishing activities.*

# DEDICATION

For my brother, William, for his love and enthusiasm for
things that are important.

       G.G.B.

For my father.

       C.D.

For those who share our passion for clean water and healthy watersheds.

       G.G.B. and C.D.

# CONTENTS

# INTRODUCTION

Impressions of nature spring to mind as we take a quiet morning walk during the early days of March. Here in the Northeast, there is still plenty of snow amid the maple, beech, and cedar trees. The spring thaw has just begun, the air is heavy with fog, and while the winter has been easy, there is enough snow left in the woods to challenge a tall pair of boots.

It is a magical time of year. In this wintry region, warm spring days appear hopelessly distant and elusive. But beneath the snow and in the rising sap of trees, the signs of reawakening and rebirth are singing loud and clear. Everywhere we look, water is flowing again!

The forest is still carpeted in deep snow, but here and there, streams are playing hide and seek. Irresistibly, we are drawn to these places where short-lived creeks appear and disappear beneath the snow. Within a month, though, these surface waterways will be gone.

Without a conscious thought, and with childlike fascination, we follow the course downstream. The first hint is a gentle depression in the snow, then a broad sheet of ice, and finally, much farther along, the first bit of open water. It's just a trickling flow – certainly no Mississippi or Fraser River here, but even those mighty rivers have modest origins. In places, the water meanders unhurried through the thawing earth. Elsewhere, there are narrow and clearly formed rivulets – perfect little streams, flowing fast and wild over gravel and sand.

We follow this elusive stream as best we can, but ultimately it seems to disappear without a trace under the snow. Perhaps it continues to flow unseen, carving out a little gorge between the snow and frozen soil. Or perhaps its course takes it on a secret underground route toward a frozen lake. Whatever its actual path might be, we know that these particular waters, and the various pollutants and nutrients that tag along, will eventually enter rivers and lakes downstream, and finally make their way to the sea.

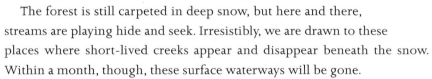

No matter where we live, we are all part of an incredible watershed story, and we are all part of nature. Water is essential to life on Earth, and it is constantly on the move. Maybe this is part of our fascination with that meltwater stream, and part of our fascination with every other stream and lake and ocean. Sooner or later, we all stare into a river and wonder where it has been, and where it is heading.

When probing the mysteries of a watershed, one can detect many things, both wondrous and worrisome. It is through such observation that we embark on a voyage of discovery and, at the same time, gain understanding and compassion. And because water is a lifeline for us all, it is the natural focus for a book like this one, which serves as an introduction to ecology and environmental issues.

In this book, we are addressing three main goals. First, we hope to promote a better understanding of the fundamentals of ecology. Because water is vital to life, we pay particular attention to water-related issues, from the simple concept of a watershed to the biological intricacies of a wetland ecosystem. (A glossary of technical terms used in the text appears on page 146.) With this ecological context in place, we aim secondly to examine and explain the numerous environmental issues affecting the health of natural ecosystems in North America. Finally, we hope that readers will become active in furthering the goals of conservation and helping to protect and restore the environment. We have included "How Can I Help?" sections throughout Part II to illustrate simple things that individuals can do to support the environment.

*Gregor Gilpin Beck and Clive Dobson*

# PART I
# ECOLOGY AND THE WATERSHED CONCEPT

At the most simplistic level, precipitation falls from sky to earth, and with that landing a never-ending cycle continues. From headwater regions, these drops of water begin their downstream journey through marshes, streams, lakes, and rivers.

It may seem strange, but there is only a finite amount of water on earth; it just keeps getting recycled over and over again. Heat from the sun causes water to evaporate into the atmosphere, and precipitation brings it back down to the earth's surface. From there, the downstream voyage within a watershed is renewed. Water may be used by plants and animals, frozen in glaciers, detained in large lakes, or evaporated back into the atmosphere. Most of it, however, eventually flows through waterways to the sea.

At one time, the face of nature was untouched by human hands, and watersheds were completely wild and wonderful systems. Countless species of plants and animals have adapted over the years to survive in almost every corner of the globe – from hot and humid to cold and dry. In the shortest blink of geological time, however, natural watersheds around the world have been dramatically altered. Increasing human populations and advancing technologies, coupled with insatiable demands for natural resources, have caused widespread environmental damage. Not a single part of the earth remains pristine. Water – our most precious of resources – suffers with each environmental insult.

The effects of forestry, agriculture, and industry, coupled with urbanization, can all be observed by listening to the voices of our waters. These voices build with each tributary stream, wetland, or spring which join together to reflect the health of the region in which they are found. For as water flows downhill from headwaters to outflows, it gathers and records the stories of environmental intrusion and improvement along its course. At the same time, it also teaches us the principles of ecology.

Arctic watersheds

Pacific watersheds

Hudson Bay

Great Lakes –
St. Lawrence
watershed

Atlantic watersheds

Mississippi River watershed

Gulf of Mexico watersheds

The Great Divide

## Watershed Definition

● A watershed is a region that drains into a particular water body. It is also called a drainage basin.

● Within large watersheds, there are many, many smaller ones; and within those smaller watersheds are even smaller ones. For example, the watersheds of the Ottawa and Richelieu Rivers are part of the larger St. Lawrence River watershed, which in turn is one of the watersheds that flow into the Atlantic Ocean.

● The Great Divide is the height of land in the Rocky Mountains that separates waters flowing west into the Pacific Ocean from those flowing east into the Gulf of Mexico or the Atlantic and Arctic Oceans.

# I

# WHAT IS
# A WATERSHED?

Whether flowing through the most remote expanses of the continents or through populous urban and rural landscapes, the earth's watercourses reflect the rocks and soils, the plants and wildlife, the fungi and microbes, and the human communities as well. Like the tranquil surface of a beaver pond at dawn, the environmental stories revealed within a watershed mirror the health of the surrounding ecosystems.

A watershed or drainage basin is a region that drains into a particular body of water, such as a river, pond, lake, or ocean. The area of land encompassed could be tiny or it could be immense. The size of a watershed, and the speed and direction of flow of its rivers, is determined by land forms. High ground, such as mountain ranges and hilltops, directs water one way or another. Within each large watershed, there are many, many smaller watersheds. For example, a small creek that flows into the Ohio River has its own watershed, but it is also part of the much larger Mississippi River watershed.

Along the edges of the major continental watersheds, the precise location where a single drop of rain lands could make a huge difference in its journey to the sea. Rain falling on the prairie of southern Saskatchewan could eventually flow into the Mississippi River and south to the Gulf of Mexico. If blown by a breeze from the south, however, the same drop of rain might land just slightly farther north, within the watershed boundaries of the South Saskatchewan River, and then flow east and north to Hudson Bay and the Arctic Ocean. No wonder important resolutions are frequently called watershed decisions.

**Height of land (watershed boundary)**

## A TYPICAL WATERSHED

- No matter where you are or where you live, you are in a watershed.
- Hills and mountains form the boundaries between watersheds, and to a large degree they also direct the path and speed of rivers.
- The upstream areas of a watershed are called the headwaters.
- As you move downhill and downstream, tiny rivulets and streams combine to form larger rivers.
- A watershed includes both water (aquatic) and land (terrestrial) components. Each watershed has a unique mixture of habitats: from brooks, rivers, and lakes to forests, farms, and even cities.
- Since the earliest days of human settlement, people have lived near rivers and lakes. Like wildlife species, people have relied on water.

**River mouth (outflow)**

Lake

Tributary stream

# Magnification of Environmental Problems

If, as individuals, we do something that is environmentally harmful, it is easy to downplay the significance of our actions by convincing ourselves that it is a minor and isolated occurrence. When one considers the immensity of the Mississippi River's entire watershed, how much damage is done by one farmer who gives cattle unlimited access to streams and rivers, thereby causing pollution and erosion? Or the apartment dweller in a large city on the Great Lakes who discards used paint thinner down the sink — how bad is it? The river seems endless . . . the lake is so large . . . surely my actions will have no real effect in the big picture.

This sentiment may make us feel better, but it is misguided and dangerous. At the local level, individual actions can and do have immediate and harmful impacts on plants and wildlife. And in the big picture, we do not live alone in our watersheds. Tens of millions of people live within the larger drainage basins of North America, and everyone contributes to pollution and ecological damage. The cumulative effects are staggering.

Consider the thousands of farms in a single county, and the hundreds of thousands in a state or province. Then think about the vast areas of land in some of the larger watersheds. Or consider the fact that a single high-rise apartment building could house over five hundred people, and a few city blocks of apartment complexes may be home to tens of thousands.

Everything we do and everything our neighbors do accumulates — from the past through the present to the future. We cannot pretend that individual actions do not matter, or that the solution to pollution is dilution. There are too many people, too many toxins, too many problems, and not enough water for that to work.

Fortunately, positive actions add up, too. When people plant trees, clean up streams, and protect wildlife, the benefits go a long way. This can help to heal past problems, reduce our impact on the environment, and restore habitats and wildlife.

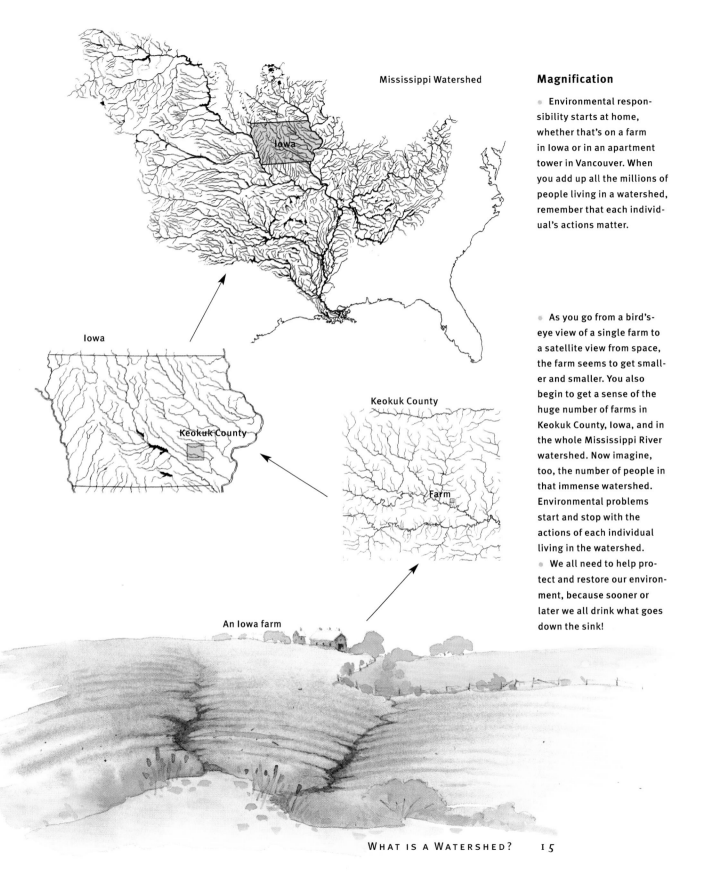

Mississippi Watershed

Iowa

Iowa

Keokuk County

Keokuk County

Farm

An Iowa farm

## Magnification

Environmental respon-
sibility starts at home,
whether that's on a farm
in Iowa or in an apartment
tower in Vancouver. When
you add up all the millions of
people living in a watershed,
remember that each individual's actions matter.

As you go from a bird's-
eye view of a single farm to
a satellite view from space,
the farm seems to get smaller and smaller. You also
begin to get a sense of the
huge number of farms in
Keokuk County, Iowa, and in
the whole Mississippi River
watershed. Now imagine,
too, the number of people in
that immense watershed.
Environmental problems
start and stop with the
actions of each individual
living in the watershed.

We all need to help pro-
tect and restore our environment, because sooner or
later we all drink what goes
down the sink!

There are many different ecological regions – or biomes – within North America, ranging from the cold, dry high Arctic to prairie grasslands and temperate coastal rainforests. These biomes are based largely on temperature and precipitation. Within each region, there are various habitats. The mood and characteristics of watersheds depend to a large degree on the biomes in which they are found.

# North American Bioregions and Their Watersheds

Traveling across North America, you can't help noticing the striking differences between regions. As you drive from east to west, forested hills and ancient, rounded mountains give way to panoramic vistas of flat or gently rolling open prairie. Continuing west, you take the roller-coaster ride over the Great Divide, through the western mountains, and down from the high country through forests and rivers to the Pacific Ocean. These dramatic changes in the landscape are obvious to any eye.

In contrast, the artificial boundaries between provinces, states, and even countries are only visible as lines on a map, as highway signs, or as international border checks. They are almost always meaningless in ecological terms.

Temperature and precipitation are the critical factors that determine the boundaries of distinct plant communities. The type of vegetation, together with climate, has an extremely important influence on which animals and other living creatures inhabit these particular regions. These different "biogeographical" regions are called biomes, bioregions, or ecoregions.

The unique characteristics of individual watersheds are determined by many factors, including topography, geology, climate, vegetation, and wildlife. Together, these characteristics shape the course and speed of rivers, the size of wetlands and lakes, and the existence of every living thing in the watershed. Some large watersheds may begin in one bioregion and end up flowing through another type of bioregion far away.

The following examples illustrate a few different types of bioregions and watersheds from various parts of North America. Here, the emphasis is on the characteristics of the surrounding land-based plant communities, large-scale geographical features, and river patterns.

# EASTERN DECIDUOUS AND MIXED FOREST BIOREGION

The eastern regions of North America, from Florida to the Great Lakes, and from the Atlantic Ocean to the prairies, are dominated by deciduous or mixed deciduous and coniferous forests. In the south, there are forests of pine; throughout Appalachia, New England, and central and eastern Canada, deciduous forests of maple, beech, birch, and oak dominate. At higher elevations and to the north, coniferous and deciduous species coexist to form a mixed temperate forest.

The land has rich soils and moderate rainfall, which can support large trees, shrubs, and smaller plants, as well as a diverse assortment of wildlife. Large watersheds, which include the Great Lakes – St. Lawrence River system and the eastern Mississippi, help ensure that soils are well drained. While much of this area remains wooded, many heavily populated regions have been cleared for intensive agriculture and urbanization.

These eastern areas were the first parts of North America colonized by European settlers. Clearing the forest for farmland was a big challenge for settlers. Once an obstacle to farming, the hardwoods and pines themselves are now an extremely valuable resource. The region remains densely populated – cities, industry, farms, and forestry are all part of the human landscape of the east.

Many of the large wildlife species of this region disappeared when the land was settled. Species such as wolves and bears are now found only in the wilder areas, and some forest birds are much less common. White-tailed deer, coyotes, and wildlife that thrives in a more open landscape are probably more numerous now than before the forests were cut.

## BOREAL FOREST AND TUNDRA BIOREGION

In northerly and mountainous regions of the continent, the landscape is shaped by the cold climate, low precipitation, occasional forest fires, and rugged topography. Compared to other North American habitats, the boreal forest (or taiga) is comparatively young, being the last region from which the glaciers retreated. This is the largest terrestrial biome on the planet, spreading across northern Asia, Europe, and North America. It also occurs much farther south, but only at high altitudes. Conifers such as black and white spruce, balsam fir, tamarack, and jack pine dominate, with a few hardy deciduous species, including poplar and birch, also present.

North of the boreal forest is the Arctic tundra. There is vegetation here, but it grows close to the ground to avoid the drying effect of strong winds. Many tree and plant species common in the boreal forest are also present above the tree line, although they are often stunted. Permanently frozen soils (or permafrost) occur beneath the tundra regions; surface soils are usually moist or wet in summer. Some other parts of the Arctic are exceedingly cold, dry, and windy, with very little soil. Few, if any, plants can survive such harsh conditions, and these areas are sometimes called polar or frozen deserts, rather than tundra.

Much of the north is relatively flat, and soils are acidic or scarce – or both. The lowlands surrounding Hudson Bay and the rugged Pre-Cambrian (or Canadian) Shield are poorly drained. Much of the available water is stagnant, boggy, or even frozen for all or most of the year. Nonetheless, northern rivers swell with spring and summer melt, and many drain vast sections of the continent.

- Human population is relatively low in the boreal forest and tundra regions. Small, scattered communities are almost completely dependent on natural resources, such as wood and mineral products and fish and game.
- The forest industry relies heavily on boreal trees to produce pulp and paper products.
- The boreal forest is home to many large mammals, including moose, wolves, and beaver. The region provides nesting habitat for a great many species of migratory birds.

## GRASSLAND BIOREGION

Sprawling across the middle of North America, and in the shadow of the Rocky and Coast Mountains to the west, lie vast open grasslands or prairie. Rainfall allows for the growth of grasses and other short herbaceous plants, but is insufficient for the establishment of large trees. Periodic fires and drought also help to maintain the grassland ecosystem.

The prairie grasslands are flat or gently rolling. Rivers carve out large valleys (or coulees), where different plant species, including large trees, can be found. These refuges provide critical water, food, and shelter for wildlife, and add to the subtle diversity and richness of the region. Grazing and burrowing animals dominate grasslands throughout the world. These species feed intensively on the vegetation, but prairie plants are adapted to survive this stress.

The grasslands of North America are drained by a relatively small number of rivers. These rivers, however, have very large drainage basins. Most of the runoff from the Canadian prairies flows into the Saskatchewan River, which eventually enters the Nelson on its way through boreal forest and tundra to Hudson Bay. The American prairies drain south through the Mississippi River into the Gulf of Mexico. Grasslands farther west drain into the Colorado, Snake, and Columbia Rivers, and eventually into the Pacific Ocean.

## WEST COAST RAINFOREST BIOREGION

While familiar with *tropical* rainforests, many people are surprised to learn that both Canada and the United States also have rainforests. Moist winds from the Pacific Ocean reach shore and are forced upwards by the Coast Mountains. The cooler temperatures of higher altitude cause rapid cooling and heavy rainfall, especially during the winter months. In summer, frequent fogs bring even more moisture to coastal forests.

From Alaska and British Columbia to northern California, these temperate rainforests are characterized by high annual precipitation of up to 160 to 240 inches (4–6 m). Vegetation varies, but is dominated by conifers, including firs, cedars, spruces, and in the south the giant redwoods. Dead organic material decomposes and is quickly absorbed by growing vegetation; consequently, soils are often relatively low in nutrients.

Because of the mountainous nature of the region, many west coast watersheds are characterized by clean and fast-flowing streams and rivers that run directly into the Pacific. The largest rivers slow only when they reach flatter terrain near the coast. Some rivers in British Columbia and southern Alaska flow into long, deep fjords which punctuate the coastline.

- North America's west coast rainforests contain a rich biodiversity, with a tremendous wealth of plants and animals. Many of these species are adapted to life in the forest or along the edge of the sea.
- The climate on the west coast of North America is rainy but very mild. Tens of millions of people now call the Pacific region home – most having settled near the mouth of a river or along the ocean.
- Forestry, farming, fishing, and industry are major economic forces in the region.
- Because of warm temperatures and plentiful rain, trees can grow quickly and to very large sizes. California redwoods can reach heights of over 325 feet (100 m).

- Deserts are characterized by their extremely dry (or arid) conditions. Relatively few species of flora and fauna can survive these conditions.
- Historically, deserts were avoided by most people, but underground springs and distant rivers now supply water for agriculture and for large cities, such as Las Vegas and Palm Springs.
- Deserts in the central and southern parts of North America are extremely hot during summer days, but cold at night and in the winter.
- Since the rains are often short-lived, plants grow and flower very quickly. With a burst in plant life, wildlife suddenly appears.

- In the far north, the Arctic winter is frigid, and summer is short-lived with temperatures rising only barely above freezing point. In many areas of the high Arctic, water is extremely scarce. These areas are sometimes called polar deserts.
- In the high Arctic and central and southern deserts, plants and wildlife are usually found close to whatever water sources are available. These oases are essential for life.

## DESERT BIOREGION

Deserts suffer the extremes. Temperatures can vary dramatically from day to night, as well as with the season. Water, though, is always very scarce, and evaporation exceeds rainfall. When rainfall does occur, it is usually infrequent and limited, and may occur in short, heavy episodes. As a result, much of the precipitation does not have time to be absorbed by the soil, but runs off through short-lived and turbulent streams and rivers or evaporates. Hot deserts are located in the central and southern parts of North America, often on the drier, lee side of mountain ranges. In some parts of the Arctic, conditions are so dry that these regions are also considered deserts, albeit cold, frozen polar ones.

Winds, rains, and occasionally torrential watercourses can cause rapid erosion. In some desert regions, water may collect in canyons and lakes; sometimes, rainfall evaporates quickly, leaving behind salty lakes and deposits. The types of vegetation and wildlife vary depending upon the amount of water, the temperature, and the nutrients available. Regardless of the species, each must have adaptations to help it find water or be able to survive until rains arrive. Where water is available, flora and fauna can thrive. These oases provide critical habitat for plants and animals.

# 2

# HOW WATERSHEDS WORK:

## Water and Nutrient Cycles

The survival of all living things is dependent upon the continuous cycling of water and nutrients through ecosystems. The same chemical elements found today within the cells of your body might yesterday have been found within the cells of an apple or orange – perhaps from a tree in another state or province. Tomorrow, these elements may still be part of your body, having been used for growth or tissue repair. Alternatively, they may be recycled yet again as nutrients for plants in a sewage lagoon or in a lake. Through an eternity, the same nutrients have been used again and again and again. It may seem crazy, but to some degree we are all reincarnated – or at least recycled – dinosaurs!

Plants and animals rely on a large number of chemical nutrients to survive. Carbon, oxygen, hydrogen, and nitrogen are the most abundant elements in living things, making up about 96 percent of all tissues. Oxygen is also required to fuel the metabolic process of most organisms. Phosphorus, sulfur, calcium, and potassium make up most of the remaining 4 percent, although numerous other elements are needed in very small amounts.

The continuous and critical reuse of chemical nutrients is made possible by the input of energy provided by the sun. This solar energy is required by plants for photosynthesis; it also causes water to evaporate, which is essential for the hydrological cycle. Nutrients are reused, but a constant input of solar energy is always needed, hence the common ecological expression that "nutrients cycle and energy flows."

As water continues to cycle on earth, many nutrients and pollutants get washed into waterways and tag along. In this way, the water cycle helps to drive some of the other nutrient and pollution cycles, too, making it particularly important. Many environmental problems relate to the way in which humans have interfered with these various cycles, so we will first take a look at how these cycles work.

- The nutrients on earth are reused over and over again.
- Energy from the sun fuels the recycling of water and other chemical nutrients.
- Nutrients cycle and energy flows.

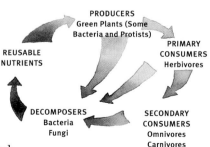

REUSABLE NUTRIENTS

PRODUCERS
Green Plants (Some Bacteria and Protists)

PRIMARY CONSUMERS
Herbivores

DECOMPOSERS
Bacteria
Fungi

SECONDARY CONSUMERS
Omnivores
Carnivores

Heat energy from the sun causes evaporation of water from the surface of lakes, rivers, and oceans, as well as from the land. Moisture evaporating from plants into the atmosphere is called "transpiration."

Whether through ground-water, surface runoff, or rivers, water eventually returns to lakes and oceans, and continues the water cycle through evaporation.

## THE WATER CYCLE

Water is constantly changing forms. It can remain frozen solid in glaciers or permafrost for thousands of years, moving slowly if at all. As a liquid or gas, it moves far more rapidly, flowing with river and ocean currents, or drifting as a vapor through wind currents. As if by magic, water is continually changing: from solid to liquid to gas and back again. The force that drives these metamorphoses is the energy from the sun.

Heat from the sun causes water to evaporate. When the air temperature gets cool enough, this water vapor eventually condenses and falls back down to earth. Because of the force of gravity, water runs downhill – over land, through the ground, or as part of rivers and streams. Some of it quickly evaporates back

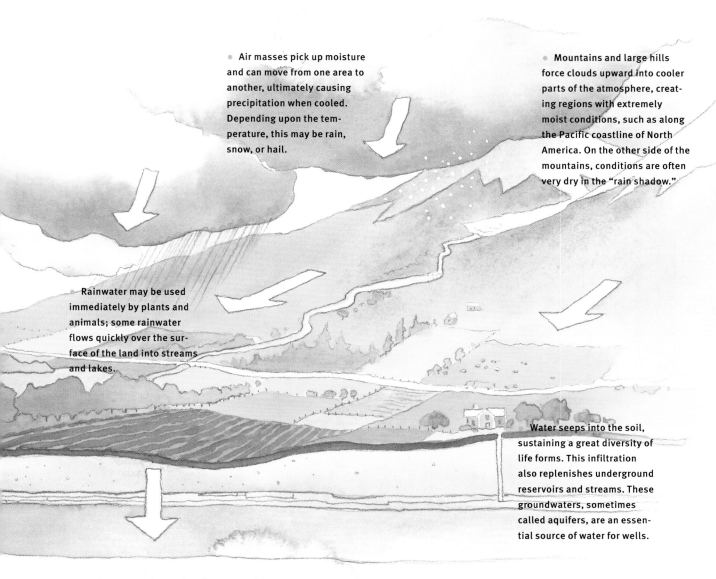

Air masses pick up moisture and can move from one area to another, ultimately causing precipitation when cooled. Depending upon the temperature, this may be rain, snow, or hail.

Mountains and large hills force clouds upward into cooler parts of the atmosphere, creating regions with extremely moist conditions, such as along the Pacific coastline of North America. On the other side of the mountains, conditions are often very dry in the "rain shadow."

Rainwater may be used immediately by plants and animals; some rainwater flows quickly over the surface of the land into streams and lakes.

Water seeps into the soil, sustaining a great diversity of life forms. This infiltration also replenishes underground reservoirs and streams. These groundwaters, sometimes called aquifers, are an essential source of water for wells.

into the air and much of it is used by plants and animals. Water is also stored for long periods of time in large lakes and the oceans.

Water, with its two molecules of hydrogen and one of oxygen ($H_2O$), does far more than provide sustenance and life to plants and animals. It helps to dissolve and transport other essential chemical nutrients, such as carbon, nitrogen, and phosphorus. Therefore, it plays an extremely important role not only in nutrient cycles, but also in the distribution and abundance of all living things. An appreciation of the intricacies of the water cycle is the foundation for understanding a wide variety of ecological and environmental concerns, and watershed issues in particular.

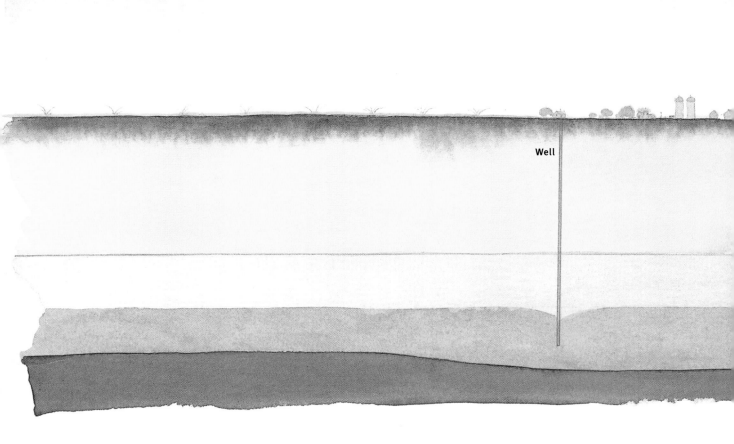

Well

## Aquifers and Groundwater

Groundwater aquifers are formed and replenished when rainwater from the surface trickles down (or percolates) through the soil. This filtering process helps to purify water. A lot of water will enter the ground if the soil is sandy and light. Less water soaks into the ground when the soil is heavily compressed or if there is a lot of clay. Water will keep traveling downward into the soil until it hits a hard, or impermeable, layer of rock or clay. The aquifer is the layer of water-saturated soil above the impermeable layer.

The term "water table" is often used to describe the depth at which groundwater can be found, but it actually refers to the uppermost limit of the aquifer. Groundwater can sometimes also be found in deeper aquifers underneath layers

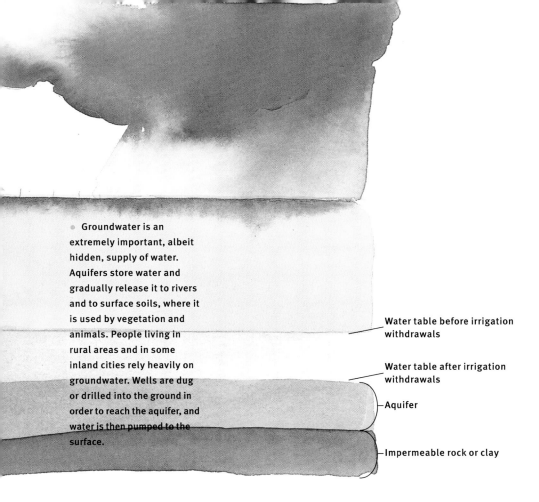

- Groundwater is an extremely important, albeit hidden, supply of water. Aquifers store water and gradually release it to rivers and to surface soils, where it is used by vegetation and animals. People living in rural areas and in some inland cities rely heavily on groundwater. Wells are dug or drilled into the ground in order to reach the aquifer, and water is then pumped to the surface.

Water table before irrigation withdrawals

Water table after irrigation withdrawals

Aquifer

Impermeable rock or clay

- Water from rainfall and snowmelt slowly percolates down through the soil to replenish groundwater.
- Underground water stores are called "aquifers" or simply groundwater.
- Impermeable layers of rock or clay block the flow of water downward, and form the bottom of aquifers.
- Wells tap into aquifers to supply water to farms and rural areas, as well as towns and cities which are located long distances from major lakes and rivers.
- Wells are dug through soil in order to reach water reserves. Sometimes, groundwater is found beneath buried layers of rock, and in these cases wells must be drilled even deeper.
- The distance between the soil surface and the top of the aquifer indicates the depth of the water table. If water is being removed faster than it is replenished, then the water table falls, and the total amount of water in the aquifer decreases.

of semipermeable rocks. The size of aquifers and the movement of underground water depends partly on the nature of the buried layers of rock and clay and partly on the amount of rainfall in the region. Natural springs are formed when water moving through the soil is forced to the surface by an impermeable layer.

Some aquifers are small, while others are very big. The largest aquifer in the world is the Ogallala. It is found under eight of the western states, from South Dakota and Wyoming in the north to New Mexico and Texas in the south. This enormous reserve was formed in prehistoric times and is now being used by cities and industry, and for intensive agriculture. The rate of water removal far exceeds the natural rate of replenishment, and the resource is being depleted.

# THE CARBON CYCLE

Carbon is the building block upon which living things are constructed. In the chemistry of life, carbon is the backbone for all the complex molecules that hold us together.

Through the process of photosynthesis, plants are able to convert carbon dioxide gas from the atmosphere into carbon-containing sugars and other solid substances. These in turn provide food and energy for animals, fungi, and innumerable microbes. So we owe a great deal of thanks to plants for getting things started. As animals go about their daily activities, carbon dioxide is produced as a metabolic waste and released back into the atmosphere through breathing and cellular respiration. Plants, too, consume self-made sugars and return some carbon dioxide to the air; when plants are burned, their tremendous store of carbon is released into the atmosphere.

- Carbon is present in the atmosphere as a gas in the form of carbon dioxide ($CO_2$).
- During photosynthesis, plants take in carbon dioxide from the atmosphere, convert it to sugars and other forms of stored food energy, and release oxygen as a by-product. Plants, together with algae and some bacteria, are called "producers." They provide stored food energy, which can be used by consumer organisms, and they are the basis of all food chains.
- In aquatic ecosystems, photosynthesis is only possible in the surface waters, which the sun can reach. Here, microscopic phytoplankton and some larger plants photosynthesize, producing oxygen and converting carbon dioxide into new plant tissues.
- Carbon and other nutrients are recycled through death, decay, and waste products, such as feces, which enrich the soil. Some nutrients also get washed into rivers, lakes, and other bodies of water.
- Over millions of years, dead organic matter can get turned into underground sources of fossil fuels, such as coal, oil, and gas.

Dead animals, fallen leaves, and all the other solid and less glamorous by-products of life contain carbon. This provides a nutrient-rich smorgasbord for creatures on land and in the water. Much of this bounty will eventually be washed into streams, rivers, lakes, and finally to the sea. All along the way, decomposition continues and carbon changes form yet again. While carbon can move back and forth between water and air, the oceans frequently hold carbon atoms captive for millions of years.

**MICRO-VIEW OF THE CARBON CYCLE**

● Plants take carbon and other needed nutrients from the air, water, and soil and change it into food for animals – from slugs to people!

● When one creature eats another, those nutrients are passed up the food chain. Herbivores feed on plants, carnivores eat animals, and omnivores eat both.

● Decomposers such as bacteria, fungi, beetles, and worms are the unsung heroes of the nutrient cycles. These organisms chemically break down fallen leaves, dead trees and animals, feces and the like, and make those nutrients available for new life. Without them, nothing would rot and nutrients would not be recycled.

● Much of the decomposition process takes place at soil level, where plant roots quickly reuse nutrients such as carbon as they become available.

## THE NITROGEN CYCLE

In many ways, all nutrient cycles are quite similar. These chemical elements are caught in a never-ending cycle — sometimes part of living things and sometimes carefree molecules drifting in the air or bobbing in the water.

Although we may not be aware of it, nitrogen is constantly in our faces. About 78 percent of the air we breathe is actually nitrogen gas. Living things require lots of nitrogen, especially for piecing together proteins. Unfortunately, plants and animals cannot just tap into the abundant nitrogen in the air because it is locked in an unusable form. To get things started, once again animals need plants.

Without doubt, the stars of the nitrogen cycle are microorganisms. Bacteria in the soil and blue-green algae (or cyanobacteria) in aquatic habitats are able to convert this atmospheric nitrogen into a form that plants can soak up in a conversion process called nitrogen fixation. Some plants, known as legumes, have colonies of bacteria living in root nodules which fix atmospheric nitrogen into a form usable by plants. They are extremely important in helping to replenish nitrogen in the soil.

After this ecological kick-start gets things rolling, nitrogen can then be passed up the food chain along with carbon and other nutrients. And what happens in the end? Well, we may not want to think about it too much, but excretion, death, and decomposition return nitrogen to soils and aquatic environments in a very handy ecological form, all ready to be reused by plants and animals once again.

Air is composed of approximately 78 percent nitrogen, 21 percent oxygen, 0.03 percent carbon dioxide, and small amounts of other natural gases and pollutants.

Lightning and radiation also convert some atmospheric nitrogen into ammonia and nitrates, but this is much less important than the role of bacteria.

Living things require nitrogen to make proteins and other biological substances. However, most plants and animals are unable to use the abundant nitrogen gas found in air.

Consumer organisms get their nitrogen when they feed – first herbivores and then carnivores in turn. Urine, feces, and the decomposition of plant and animal tissue return nitrogen to the ecosystem in a form that can be quickly reused by plants.

Nitrogen-fixing bacteria in soil and water convert nitrogen gas into nitrates and ammonia, in a chemical form plants can use. Some plants, such as clover, have colonies of these bacteria living in round nodules attached to their roots.

A lot of nitrogen from the land gets washed into rivers, lakes, and oceans. Some of this is from natural sources, and some is pollution from excess fertilizers, livestock, and urban sources.

# THE SULFUR CYCLE

Sulfur alternates between solid and gas forms as it cycles in the environment, similar in this way to the nitrogen cycle. Most of the earth's sulfur is held fast in a concrete, material form – either as part of living or of nonliving things. Natural weathering of rocks releases some sulfur minerals directly into soil and water. Volcanic eruptions, the decomposition process, the burning of fossil fuels, and the processing of metals all release sulfur into the atmosphere, and they share that telltale rotten-egg smell.

The decomposition of organic wastes and dead organisms provides a major source of this essential element. Sulfur in the atmosphere, regardless of whether it is natural or from pollution, eventually returns to the earth's surfaces. Plants need a small amount of sulfur and obtain it through their roots. Animals, in turn, get their required dose through the foods they eat.

Acidic precipitation, commonly called "acid rain," occurs when there is an excess of sulfur and nitrogen gases in the atmosphere from pollution. This major environmental problem can kill or harm both plant and animal life. In some locations, the weathering of certain types of rocks can release large amounts of these pollutants, which is also extremely damaging.

● Volcanic activity releases sulfur to the atmosphere in the form of sulfur dioxide and hydrogen sulfide.

The burning of coal and other fossil fuels also produces sulfur dioxide, which combines with moisture in the atmosphere to form acid rain. Industry, electric utilities, and motor vehicles are major contributors to acid rain.

● Decomposition releases sulfur and other nutrients, making them available for new plant and animal life. Like nitrogen, sulfur is essential for the formation of proteins.

● During the decomposition process, bacteria produce hydrogen sulfide and sulfate gases. This process occurs naturally in marshes, ponds, lakes, and oceans, as well as in soil. Hydrogen sulfide has the characteristic odor of rotten eggs and is sometimes called "swamp gas."

# THE PHOSPHORUS CYCLE

Phosphorus is another nutrient essential to living things. Unlike nitrogen, phosphorus is not found in the atmosphere, and it is fairly scarce in most natural ecosystems. Fortunately, a little goes a long way toward building proteins and cell membranes and supplying energy to individual cells.

Phosphorus dissolves from rocks and minerals, and becomes available to plants for growth. In turn, it is passed along to herbivore and then carnivore consumers, ultimately being returned to the soil by fungi and bacteria through the process of decomposition. For the most part, all of this cycling occurs on a pretty local scale, with little long-distance movement of phosphorus. Some phosphorus, however, does enter watercourses and feeds aquatic plants; it can also settle to the bottom and become part of newly forming rocks.

• Guano from seabird droppings is rich in phosphorus. In some coastal areas, it has been mined as a natural source of fertilizer.

**Most household and agricultural plant fertilizers contain nitrogen, phosphorus, and potassium. Frequently, rain and spring runoff wash fertilizer from farms into streams and lakes.**

**This overabundance of nutrients, especially phosphorus, causes algae to grow and reproduce extremely fast. In many places, waterways have been choked by these algae blooms, ultimately reducing the amount of oxygen and aquatic life.**

• Phosphorus is an essential nutrient for plant growth and survival, but it is needed only in very small amounts.
• In ecosystems undisturbed by human activity, phosphorus may be in short supply, limiting plant growth.
• Phosphorus is found in rocks, soil, water, and living things, but is not present in the atmosphere.

# 3
# FROM HEADWATERS
# TO OUTFLOWS:
## Parts of a Watershed

Tributary stream

Dam and reservoir

From the moment a drop of rain hits the ground, it begins a watershed odyssey. That one drop may travel as surface runoff, entering a small stream, then a river, eventually making its way to a lake or the sea. It may flow as groundwater, or perhaps fall directly into a lake. At some point, it will probably make its way through a wetland of one form or another.

The downstream trip may be very short or extraordinarily long. This depends mostly upon the topography, or drainage pattern, of the land. In a region where water runs from the mountainside straight into the sea, a micro-watershed could be measured in hundreds of feet or meters. In marked contrast, a longer trip could begin in the wooded Cypress Hills of Alberta and Saskatchewan, or with a tumble down from the Rocky Mountains. The ultimate destination could be more than 2,500 miles (4,000 km) away in the Gulf of Mexico. No question, this one is an odyssey: a scenic tour of two Canadian provinces and over a dozen American states.

Regardless of length and location, watersheds share many common themes and characteristics. They all flow down from headwaters to outflows, usually into the ocean. And from the original source to the ultimate sink or destination, water passes through numerous component parts. Fast-running streams and rivers are very different ecologically from comparatively stable, slow-moving lake systems. Mysterious, brooding bogs and sedge marshes are critical parts of

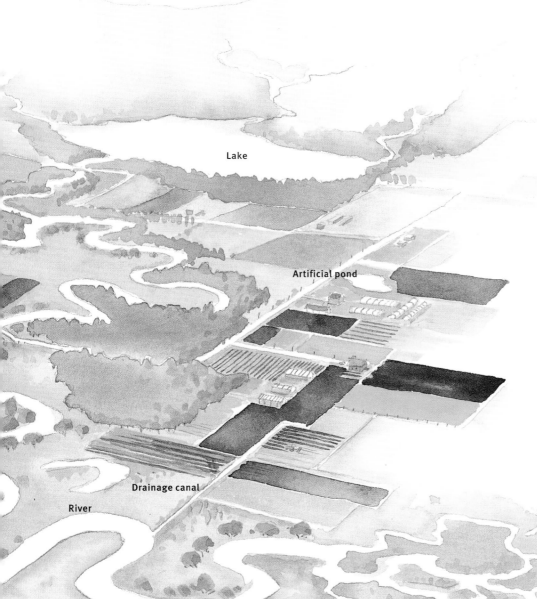

Lake

Artificial pond

Drainage canal

River

Estuary

watersheds, too, but attract very different wildlife – and people. At the downstream end of the system – at the outflow where a river meets the sea – pulsing
tides plus abundant water and nutrients make estuaries the dynamic ecological
climax to a watershed odyssey.

Like pieces of a favorite jigsaw puzzle, each part of a watershed is unique and
essential. If one piece is missing or damaged, the result compromises the whole
picture. Therefore, we must understand and protect not only individual places
and species, but complete ecological systems. This is particularly important with
watersheds, where the effects of what goes on upstream become magnified and
increasingly complicated downstream.

# Wetlands

For all too many of us, wetlands are a dark and shadowy world, home to leeches, mosquitoes, and the like – the unseemly underbelly of an otherwise upbeat aquatic system of frolicking streams and smiling ponds. But wetlands are the unwavering workhorses of the system, serving vital ecological roles that maintain the environmental health of the entire watershed.

There are many different types of wetlands, but they share the common characteristic of being saturated with water all or at least part of the time, often because of their relatively flat topography. The vegetation and soils of wetlands mean that they can act as tremendous sponges, absorbing water from rainstorms and snowmelt, then releasing it gradually to provide streams and groundwater with a steady supply of fresh water. Not only does this mean a more reliable source of water for plants and animals, but it also greatly reduces the chances of floods downstream. Rivers help to carry water away relatively quickly, whereas wetlands help to keep it in one place longer.

Wetland plants act as natural filters. Marsh plants are particularly effective at soaking up large quantities of excess nitrogen and phosphorus, reducing water pollution downstream. They can also absorb heavy metals and other pollutants, which further helps to purify water.

And of utmost importance, marshes, swamps, and bogs all provide essential habitat for an astonishing diversity of plants and wildlife. For example, small and marshy "pothole" sloughs dot the prairies and are the single most important nesting area for ducks on the continent. Weedy, wetland edges of lakes and ponds provide protection and spawning grounds for fish and their prey. Hollow swamp trees provide a safe home to flying squirrels and swallows. And rare orchids flower in the privacy of bogs.

## MARSHES

Marshes are shallow water wetlands (less than 3 feet or 1 meter in depth) which usually remain wet, but not stagnant, throughout the year. They are common along the edges of rivers, lakes, ponds, and the sea, and in other low-lying areas. Some marshes contain fresh water whereas others are salt water.

A rich diversity of emergent plants grows in marshes, including cattails, bulrushes, arrowhead, reeds, pickerel weed, and some grasses, rushes, and sedges. These are plants which have their roots and lower stems in the rich, wet soil and their upper foliage in the air. Marsh plants receive an ample supply of water and nutrients from below, and sunlight and carbon dioxide from the air above — all the ingredients for photosynthesis. Marshes are some of the most ecologically productive regions in the world, meaning that the rate of photosynthesis is very high and that plant growth is very fast and abundant.

Life of all sorts abounds — the food web of a marsh includes an inspiring diversity. The complicated shallows, rich with colorful greens and browns, contain food and hiding places for creatures great and small. Marshes are a natural nursery — fish, ducks, frogs, and insects, to name a few, seek these places to raise their young, and predators follow.

- Wetlands are the workhorses of the aquatic system, serving vital ecological roles that maintain the environmental health of the entire watershed.
- Wetlands can occur along the edges of rivers, streams, lakes, and saltwater inlets, as well as in other areas where there is a depression in the land.

## BOGS

Bogs occur in poorly drained freshwater regions, and are particularly common in the boreal forest and tundra regions of the north.

They are waterlogged with stagnant, brown, acidic water that contains little or no dissolved oxygen. Dead plant and animal matter does not decompose fully, so few raw nutrients are available for new plant growth.

Sphagnum moss, black spruce, and tamarack can tolerate the harsh bog conditions, and are the dominant plant species. Often, they grow as a floating mat of vegetation along the water's edge. Occasionally, they will even form floating "islands" in the middle of a large bog.

The partially decomposed moss and plant material in bogs is called peat. It is frequently "mined" and sold in stores as peat moss for use in gardens, where it conditions the soil and serves as a natural fertilizer. The peat quickly decomposes when exposed to oxygen, water, and decomposer organisms in soil.

Some plant species, such as sundews, pitcher plants, and the Venus fly trap, are carnivorous. These fascinating plants supplement their diet by trapping insects. The dead insects provide extra nutrients, especially nitrogen, which helps the plants survive.

Over the centuries, bogs gradually become drier and fill in with peat. This allows for the growth of different plant species and larger trees.

# FENS

Fens, another type of peat-containing wetland, are dominated by grasses, sedges, and some moss species, but not sphagnum, and resemble a flooded field of hay. Unlike bogs, there is at least a moderate – albeit slow – flow of fresh water through the fen.

The bottom of a fen is often limestone, so fen water is usually alkaline, as opposed to acidic, and nutrients may be a little more abundant than in bogs. Oxygen levels, however, are low and the scarcity of bacteria means that decomposition occurs only slowly.

Fens are saturated, and hold huge volumes of water. Therefore, they are extremely important areas for water filtration and storage, and they help to reduce the risk of flooding. Vegetation and organic matter in a fen usually go from the surface all the way down to bedrock, whereas in a bog, vegetation often floats above open water.

## Swamps

Swamps are wetlands with open surface waters. Some trees and large shrubs grow in swamps, but the wet conditions are not tolerated by many species. In the northern part of the continent, silver maple, red maple, cedar, alder, and willows may all be found growing in swampy conditions. In the south, species such as bald cypress, water oak, and swamp black gum are found in swamps. Most swamps are freshwater, but the mangrove swamps of Florida are flooded with salt water during high tides.

**BALD CYPRESS SWAMP**

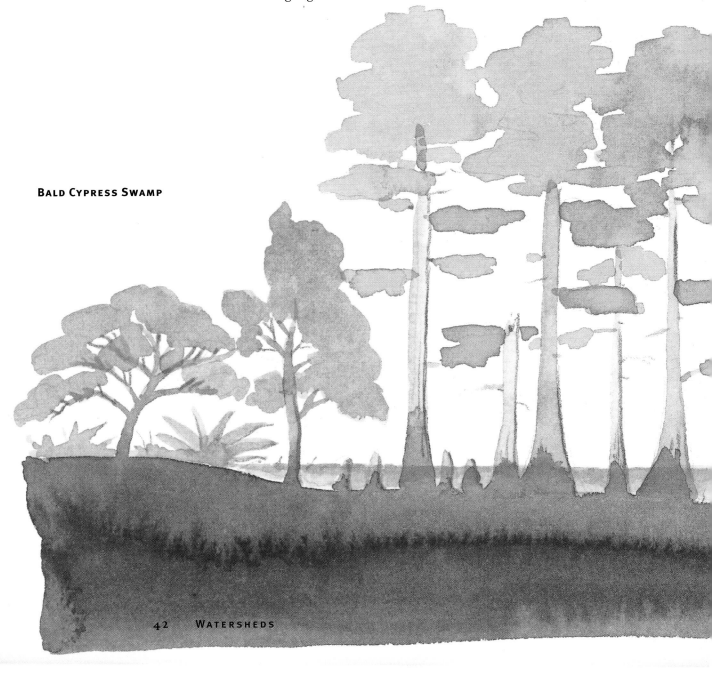

Swamps have the reputation of being scary and mysterious. There really is something mysterious about paddling around in what is essentially a swampy, sunken forest – filled with both living and dead trees, and a lot of strange noises!

Land creatures must be able to move from tree to tree within the swamp. Some species, like snakes, will swim, and many others can fly. One species of mammal – the flying squirrel – has wing-like flaps of skin between its front and back limbs which allows it to glide from tree to tree. Many predators avoid water, which makes swamps a somewhat safer habitat for smaller animals.

When the trees in a swamp die, they provide important habitat for many creatures. Woodpeckers feed on the insects inside the wood and drill holes in the trunk for nesting cavities. Swallows, flying squirrels, and even some types of ducks later use these excavations to raise their young as well. Colonies of great blue herons build nests high in the secrecy and protection of tall swamp trees.

# The Flowing Waters of Streams and Rivers

Streams and rivers are waters on the move, and each has a story to tell. Whether one is exploring by canoe or wading up a babbling brook, one cannot help but wonder what is around the next bend or up the various tributaries. These natural corridors have been major travel and trade routes for humans from ancient to modern times.

Flowing water gurgles over riffles, churns white and foamy through rapids, and cascades over a waterfall. Through this action, oxygen from the air dissolves into the water. This exchange of oxygen means life for many aquatic organisms, both in the stream and in ponds and lakes downstream. In turn, these organisms provide dinner for terrestrial wildlife. Some animals, such as trout and salmon, require clean, cool streams to survive and to reproduce in, and they also need water with high levels of dissolved oxygen. While some wildlife species are

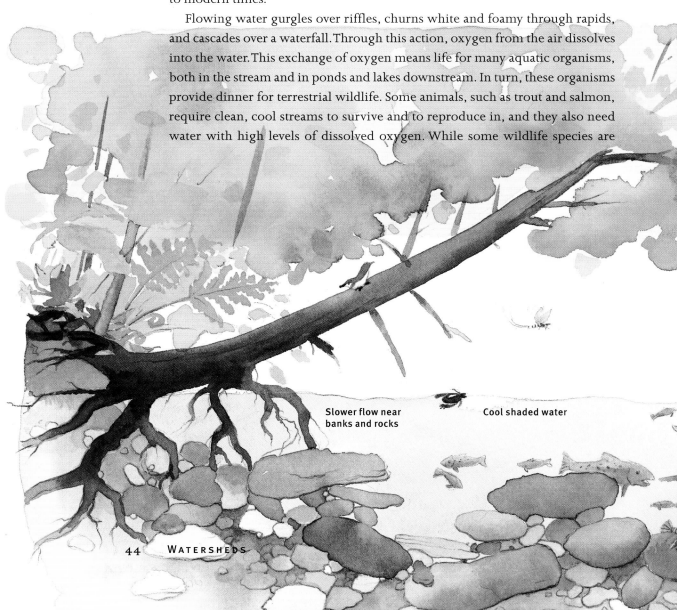

Slower flow near
banks and rocks

Cool shaded water

adapted to living in oxygen-poor water, most need moderate to high levels of dissolved oxygen.

Rivers, streams, and brooks share the common feature of having flowing water, but their individual characteristics vary tremendously, depending upon the size of the watershed and where each watercourse is located in North America. They differ greatly in speed, temperature, clarity, and the nature of their banks and bottoms. All of these factors affect the types of wildlife to be found, and dictate which adaptations are most necessary for survival.

Small tributary streams are often clear and clean as they tumble down from mountains and hillsides. Large rivers, such as the Mississippi, Saskatchewan, or Hudson, become slower as they reach flatter ground farther downstream. They pick up sediment from riverbanks and erosion, which makes them muddier.

Organisms found in fast-flowing water have special adaptations which help them live in the current. Most river fish are very streamlined and are strong swimmers.

Smaller creatures, especially aquatic insects and invertebrates, live among the rocks on the bottom and edges, where the current is much slower – even in a very large, fast river. They often have flattened bodies to reduce the risk of being washed away. Some other types will burrow into the bottom sediments and live there.

Fast-flowing waterways are usually very rich in oxygen, which is an essential feature for many fish and other animals. Oxygen dissolves into water as it tumbles over rocks in rapids, or crashes over waterfalls.

Cold water can "hold" more oxygen than warm water. Trees overhanging a stream cool the water and help keep the oxygen levels high – this keeps trout, salmon, and other oxygen-lovers happy.

Riffle

Pool

- Rivers and streams alternate between fast- and slow-moving segments. Riffles are relatively shallow and fast, with rocky or gravel bottoms, whereas pools are deeper, slower, and often have a muddy or sandy bottom.

Anglers know that many fish rest and feed in pools as they migrate upstream. Whitewater canoeists, kayakers, and rafters, too, know that after navigating a big rapid (a "monster riffle") they will be able to relax for a bit in the inevitable pool of calmer water farther along.

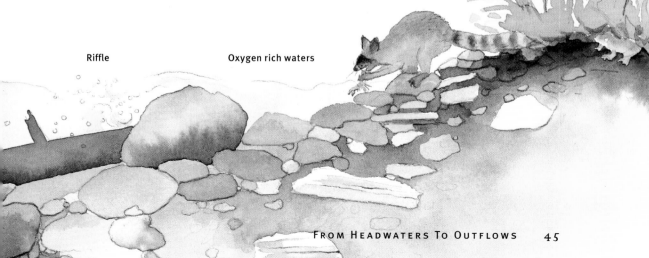

Riffle                    Oxygen rich waters

# The Still Waters of Ponds and Lakes

Lakes beckon us to their shores to enjoy swimming, boating, and camping. But despite years of recreational pursuits, what do most of us really know about a lake? Sure, we engage in casual biological studies from the dry end of a fishing pole, or keep tabs on loons and herons, but what's going on below the surface?

Basically, lakes and ponds are big holes in the ground, filled with standing water supplied by the local watershed. Things are, of course, far more interesting and complex than that. Life in the shallows is very different from that found in the depths. Some fish and wildlife swim freely within the water column,

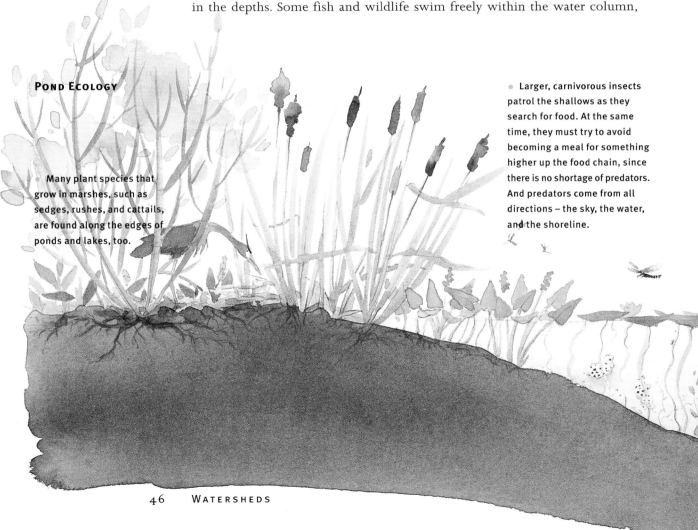

## POND ECOLOGY

Many plant species that grow in marshes, such as sedges, rushes, and cattails, are found along the edges of ponds and lakes, too.

Larger, carnivorous insects patrol the shallows as they search for food. At the same time, they must try to avoid becoming a meal for something higher up the food chain, since there is no shortage of predators. And predators come from all directions – the sky, the water, and the shoreline.

which is anywhere from the lake's bottom to the surface, whereas other species prefer to crawl around on rocks or plants.

Individual water bodies have individual characteristics. Some lakes are surrounded by fertile lands and have lots of nutrients and life; these are "eutrophic" lakes. Other lakes are "oligotrophic"; these have fewer nutrients, are deeper and colder, and are usually found in more northerly or mountainous areas. Eutrophic lakes are often murkier than oligotrophic ones because the greater amount of nutrients encourages more growth of algae and phytoplankton.

The difference between a lake and a pond is in the depth and not the size. In a pond, sunlight reaches all the way to the bottom, whereas in a lake the light does not reach the bottom. It is possible, therefore, for a pond to be larger in surface area than a lake. The actual depth to which light will penetrate depends on the clarity, or amount of suspended matter in the water. The lack of light at the bottom of lakes means that plants do not grow there. This in turn affects the distribution of plant-eating organisms and also carnivores.

As they tumble downhill, rivers and streams absorb oxygen from the atmosphere. In lakes and ponds, only surface waters can absorb oxygen directly from the air, helped along by wave action. Sunny, shallow waters also get oxygen from both large and microscopic plants. This is fine for life in ponds and near the surface of lakes, but not for the murky depths.

The abundance and diversity of life in shallow wetlands, such as ponds, is astounding. If you sit quietly beside a pond for a few moments on a sunny day and watch just one spot, you are sure to see all sorts of things creeping and crawling. And maybe something much larger, like an osprey or a racoon or muskrat.

Pondweed and water lilies are especially suited to growing in the deeper parts of a pond. They have long stems, which bring their leaves to the surface where sunlight is brightest.

Pond vegetation provides a diverse habitat for aquatic creatures. Herbivorous animals, from ducks to snails, feed actively on the leaves and also on the algae that grows on the plants.

In a pond, sunlight reaches all the way to the bottom and nutrients are usually plentiful. As a result, large plants can be found growing throughout, in addition to the microscopic algae, which are free-floating (or planktonic).

# LAKE ECOLOGY

In summer, most surface waters have ample oxygen and are warm, which is great for swimmers. But this layer does not mix with colder, deeper waters. Down below, mysterious creatures live and organic debris rots, consuming precious oxygen. By late summer, oxygen is in very short supply in the abyss: bad news for trout and other oxygen-dependent creatures. As autumn approaches, lakes cool and the temperature becomes more uniform throughout. Surface and deep waters can then mix, circulating oxygen and nutrients throughout – a process that occurs again in spring.

The shallow water areas along the edges of a lake can be quite similar to ponds, marshes, or other wetlands. This is the littoral zone, where sunlight reaches the bottom and plants of all sizes are found. Some plants can also grow in shallow water reefs in the middle of a lake, provided there is enough sand or mud for roots.

In the deeper parts of a lake, sunlight reaches only the upper waters, and large, rooted plants are nowhere to be found. Floating near the surface are microscopic algae (phytoplankton) and cyanobacteria, which form the basis of the open water (or limnetic) food chain. These minuscule organisms produce oxygen and provide food for zooplankton, which in turn provide food for fish and most of the other lake creatures.

## Summer

- The surface waters (or epilimnion) of a lake often become quite warm in the summer months. In contrast, the deep waters (or hypolimnion) remain cold. Separating these two zones is the thermocline (or metalimnion), where temperature changes dramatically within a few yards (about one half °F per foot, or 1°C per meter!).

- The epilimnion and hypolimnion are two very distinct regions within the lake, especially in summer.

There is very little mixing of waters between the two, so it is much like having two lakes in one – a warm lake sitting on top of a cold one.

- Despite the continuous use of oxygen by living organisms through respiration, the epilimnion maintains high oxygen levels because of photosynthesis and the mixing of oxygen from the air.

- By late summer, oxygen concentration in the depths (where it is too dark for photosynthesis) becomes very low. Fish, zooplankton, and other animals have used much of the oxygen supply. In addition, dead plant and animal materials accumulate on the bottom of lakes, and many of the decomposer microbes consume oxygen also.

- It is possible for oxygen to become completely or almost completely depleted, which would cause the death of everything except anaerobic bacteria. This usually happens only in heavily polluted or eutrophic lakes. Fortunately, the seasonal lake overturns help to restore oxygen to the lake bottoms.

- Because surface and deep waters do not mix in summer, oxygen in surface waters cannot help creatures at the bottom of the lake. Similarly, nutrients from decomposition on the bottom are not available for phytoplankton near the surface, which can limit their growth.

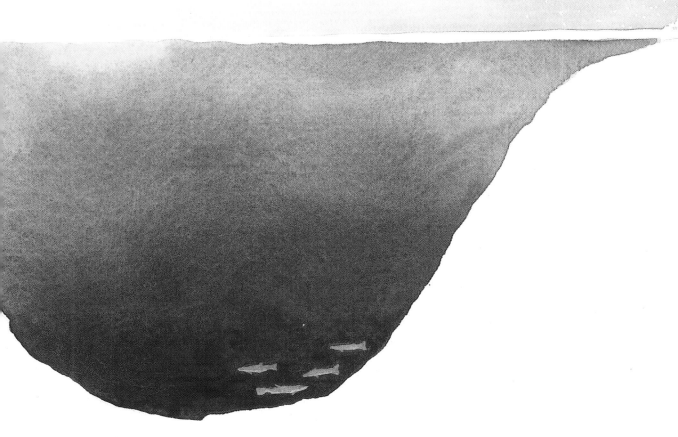

### Spring/Fall

- Surface waters cool in the fall, and the lake temperature becomes more uniform throughout. Since water is densest at 40°F (4°C), cooling epilimnion waters sink toward the bottom and force deep waters upward. With the temperature stratification gone, wind action and currents help to circulate water.

- This seasonal mixing, or overturn, brings an essential supply of oxygen to the bottom of the lake and nutrients from decomposition to the surface. During overturns, temperature, oxygen, and nutrient levels are fairly uniform at all depths.

### Winter

- During the winter, the surface temperature of lakes is very cold – so cold, in fact, that many lakes in northern regions freeze. In contrast, the temperature in the hypolimnion is usually about 40°F (4°C), a few degrees warmer than what is above it in the water column. This temperature profile is the opposite of what occurs during the summer months.

- Because of continuing decomposition by bacteria and other organisms, oxygen concentrations may decrease near the bottom over the winter.

- As the weather warms in spring, the surface water gradually reaches the critical 40°F (4°C) point. This again leads to the mixing of surface and deep waters, and ensures that the entire lake is well supplied with both oxygen and nutrients.

- The fact that ice floats is really quite remarkable. Water is the only substance for which the solid form is *less* dense than the liquid.

# Estuaries: Where Rivers Meet the Sea

At the downstream end of the watershed system, a river ultimately unites with the sea. Through history, both wildlife and people have flocked to these natural junctions. Estuaries are gateways to rivers, which lead into the hearts of continents. And, they are the main corridor through which vital nutrients leave the land and the watershed to enter marine ecosystems.

Estuaries come in all shapes and sizes. Some have rocky, well-defined shorelines, which drop steeply to deep waters. In others, the mouth of the river is shallow and the delta expansive, with abundant eel grasses and kelps. Some

rivers, such as the Mississippi, carry so much silt that the extensive deposits extend far out to sea. These deposits may then be carried along the coastline by ocean currents and spread out to form a "bird's-foot delta."

Marine plants and phytoplankton require an abundant supply of nutrients in relatively bright, shallow waters. Estuaries provide these essentials of light and food – and consequently life of all sorts flourishes. These are critical nursery areas for young fishes and invertebrates, and important feeding areas for migrating birds. In contrast, life is scarce in the clear surface waters of the open ocean. There is plenty of light, but few nutrients and little ecological productivity. The clear blue water may be beautiful to look at, but it indicates the relative scarcity of plant and animal life.

Everything that lives in an estuary must be adapted to constant change, which means plants and animals must have numerous survival strategies. In the sea, water levels rise and fall twice a day, posing difficulties for life in the shallows. The height of the tide varies from day to day, and from region to region. Water temperature and salt concentration (or salinity) also vary, depending upon tides, river currents, and other factors. In fact, in estuaries there is often a layer of lighter fresh water floating on top of denser salt water.

● An estuary occurs where a river empties into a partially enclosed area of the sea. A layer of fresh water may flow over top the cooler, saltier sea water below. Nutrients carried downstream by the river help make estuaries very rich ecologically.

● Estuaries and other coastal wetlands are among the most productive ecosystems in the world, supplying essential oxygen and food for other living things.

## ROCKY ESTUARIES

If a river enters the ocean where it is deep or where there are strong currents, most of the sediments may be quickly washed away from the river mouth. As a result, the distinction between river and ocean shoreline is very clear.

Even if delta deposits do not form in the estuary, there is an abundance of marine life. Nutrients transported by the river are present near the surface, where phytoplankton and seaweeds thrive. Zooplankton and other herbivores graze on these producers, and in turn are eaten by other marine organisms.

One such estuary occurs in Quebec where the Saguenay River meets the St. Lawrence River, in an area further enriched by the upwelling of nutrient-laden ocean currents. It is a dynamic and impressive ecosystem, where the ecological productivity yields huge numbers of a small fish called capelin. Their presence explains the local abundance of seals and whales, including the endangered population of beluga whales.

## SALTWATER MARSHES

Salt marshes are somewhat similar in appearance to their freshwater counterparts, but for the most part they are dominated by a variety of grasses. Different species grow in specific parts of the marsh, with distinct zones visible as one moves out from the shoreline.

Salt marsh plants must be able to adapt to constant change in water and nutrient availability, as well as variations in salinity. The twice daily rise and fall of the tide and the influx of fresh water from rivers makes tolerance an essential characteristic.

Many creatures live in salt marshes permanently, either among the vegetation or in the wet soils, whereas others, such as waterfowl, visit during migration or for one part of their life only.

## EEL GRASS BEDS

Dense beds of eel grass grow in the shallow, muddy waters of some estuaries. These occur most commonly in areas where they will remain under water at low tide.

The thick mats of vegetation provide an excellent place for creatures to feed and hide. As with saltwater marshes, many species of fish and invertebrates live in eel grass beds. These nursery areas are needed to protect the small and young organisms from predators. Similarly, seaweed beds (kelp, or marine algae) provide essential protection for creatures on rocky coastlines.

# 4

# Natural Changes Within Watersheds

People often assume that everything in nature remains more or less the same. Even the expression "the balance of nature" seems to suggest that natural things stay fairly constant. As a result, we assume that ecological changes are always bad and always caused by humans. This is an understandable conclusion, considering that humans have indeed imposed countless changes upon the earth, of which a great many have had terrible environmental consequences.

We must remember, though, that nature itself is always changing. While some species are better able to tolerate major environmental changes than others, all organisms must have at least some adaptations or behaviors that help them to survive small fluctuations in their living conditions. Regardless of the cause, some species will benefit from a change in the environment while others will suffer.

Natural ecological change includes major events such as the movement of glaciers or the rise and fall of sea levels during prehistoric times. It also includes major climate-related conditions, such as drought, flood, storm, and fire. Even living plants and animals cause changes in nature. Insect infestations, flooding caused by a beaver dam, and shade from a maple tree all influence the species found in a particular area.

## BEAVER PONDS AND MEADOWS

Like humans, beavers have the ability to alter their surroundings to suit their needs. When these large rodents build a dam, they can turn a small stream valley into a very large pond, measuring a few meters in depth – which is rather a major ecological change! This is certainly not good for the drowned trees and plants or the land creatures forced to relocate, but it does provide new habitat for wildlife that requires an aquatic habitat.

Beaver ponds frequently have a lot of standing dead trees in them, but eventually they rot and fall down. When the pond is no longer being used by beavers, the dam will not be maintained and breaks apart. This leaves a stream flowing through the middle of the former pond, an area now extremely rich in decomposed organic matter.

A meadow community will develop quickly, which is relatively uncommon in the northern forests where beaver are most abundant. Gradually, natural succession advances, larger plants grow, and the forest returns to complete the cycle.

# Natural Succession of Plant Communities

Forest fires, wind storms, and insect infestations can have a large impact on terrestrial ecosystems. The effect of any of these disturbances is much the same: the loss of tree cover allows more sunlight to reach the ground. This means that different plants can now grow there. Over time, the ecosystem changes as larger and different species of plants and trees become established. Eventually, a mature and more stable plant community – called a "climax community" – becomes established. This process of change is called natural succession.

If the disturbances are severe, virtually all vegetation above ground can be destroyed. Quite often, though, some trees and shrubs do survive, and new shoots and leaves grow back. The effect of these disturbances has many benefits too, since fallen trees and burnt wood return nutrients to the soil. Many plants and animals prefer the ecological conditions that follow fire or insect infestation. In fact, some ecosystems require occasional natural disturbances. For example, in the dry central plains of North America, fire limits the growth of trees and maintains the grassland community.

The biggest benefit of natural disturbances for many plants is the increase in the amount of sunlight reaching ground level. This favors the establishment of meadow plants, such as grasses and fireweed. Many of these first colonizers have airborne seeds, which travel by the thousands from other regions. Gradually, small shrubs and sun-loving trees become established, with seeds often being distributed by birds. To be successful in the competition for light, many plants grow extremely fast. Poplar trees, for example, produce lots of wind-distributed

seeds; once rooted, they often grow 6 feet or more (several meters) in a single season. If not damaged during a fire, they will also sprout innumerable shoots (or suckers) from their roots.

Once the first generation of shrubs and trees becomes established, the ground is shaded and ecological change is inevitable. Species that need full sunlight can no longer survive, but shade-tolerant species thrive and spread. The young saplings of climax species can tolerate shade for several years, but often require brighter conditions to reach maturity. This can occur when one of the trees above dies and sunlight reaches the growing sapling. The final successional stage occurs with the establishment of a climax forest or community.

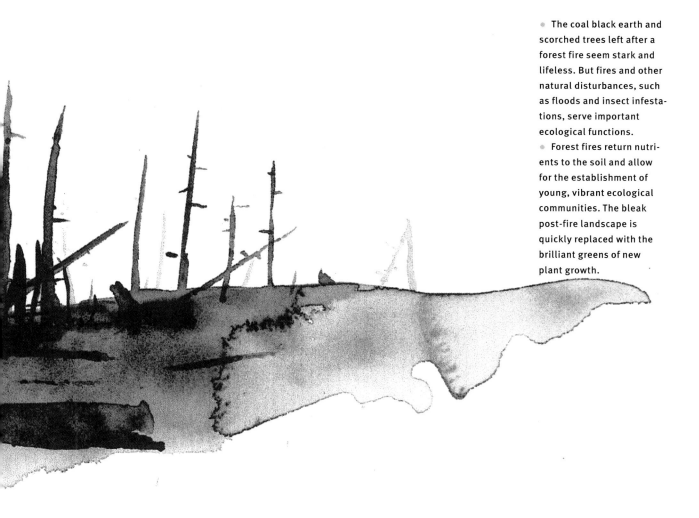

● The coal black earth and scorched trees left after a forest fire seem stark and lifeless. But fires and other natural disturbances, such as floods and insect infestations, serve important ecological functions.

● Forest fires return nutrients to the soil and allow for the establishment of young, vibrant ecological communities. The bleak post-fire landscape is quickly replaced with the brilliant greens of new plant growth.

## FLOODS

Flooding is a natural occurrence in the life of a river. Flooding occurs after very heavy rainfalls and during the spring snowmelt. As a result, river valleys (or flood plains) and estuaries are nourished with essential nutrients carried by the river.

Forests and wetlands play a major role in limiting the scale and frequency of floods. Plants use vast quantities of water, and vegetation and soils act as great sponges to slow down and reduce the amount of rainwater that enters a river. The more forests and wetlands there are, the less the chance of an unnaturally big flood.

The size and frequency of flooding in many rivers has been greatly increased in recent centuries by human activities. The loss of wetlands, the removal of vast areas of forest (especially those close to rivers), and the increase of hard surface runoff from urban areas all play a part in creating major disasters.

A great many communities have developed over the millennia in river valleys, close to fertile soils and highly productive farming land. It is both ironic and tragic that permanent settlements became established in places where the natural conditions that encouraged settlement also frequently cause so much harm.

# EUTROPHICATION OF PONDS AND SMALL LAKES

Natural change can sometimes produce surprising or even startling results. Take, for example, the mystery of the disappearing pond or small lake.

Nutrient-rich lakes and ponds support an abundance of plant and animal life. Once dead, organic matter accumulates on the bottom. Runoff from surrounding regions adds more nutrients to the lake, and soil builds up on the bottom. Gradually, over many years, the depth of the water decreases as the amount of decomposing matter increases. Fast-growing plants flourish in the shallower water. When they die, they contribute to the ever-increasing soil layer. Eventually, dry-land species of plants will be able to grow, and the pond or lake disappears — completely filled with rich soil and decomposing plants and animals.

Then the process of natural succession continues as one type of terrestrial plant community replaces another, until finally the climax stage is reached.

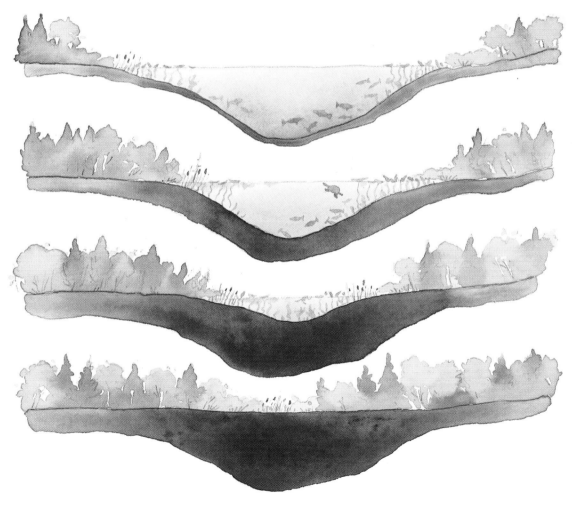

# PART II

# ENVIRONMENTAL ISSUES, IMPLICATIONS, AND SOLUTIONS

In Part 1, we introduced many fundamental principles of ecology. We looked at different North American bioregions, examined the water and nutrient cycles, and investigated the different parts of watersheds and how they function. We also saw that change can be a natural part of the watershed story.

We are all well aware, however, that human beings have also caused changes within nature. These changes are so numerous and so far reaching that it would be impossible to count them all or even to attempt to accurately assess the environmental costs. The damage caused by people has affected not only individual plants and animals, but also entire species, ecosystems, and watersheds.

Perhaps the most unbelievable thing about all this ecological damage is the speed at which it has occurred. The earth is about 4.6 billion years old. The earliest human ancestors date back about 3 million years or so, but modern civilizations are only a few thousand years old. Since then, *Homo sapiens* (Latin for "wise man") has used the ability to make tools and communicate to spread across the entire globe, even expanding out into space.

Human population has increased exponentially, especially during the last century or two. Global population may exceed 7 billion by the year 2010, according to estimates by the United Nations Population Division. And industrialization, which began only a century ago, has meant that modern societies consume natural resources at an unprecedented rate. The root cause of the majority of our environmental problems can be traced to overpopulation and industrialization.

In the following chapters, we examine some of the major environmental problems affecting North America. These include air and water pollution, and the effects of nonnative species and habitat loss on natural systems. We emphasize what is being done and what each one of us can do to improve the situation.

# 5
# AIR POLLUTION

Air pollution is a complex environmental issue. Many pollutants are transported by wind currents from one place to another or end up high in the atmosphere. This means that the pollutant can become a major concern for other watersheds, too. In many cases, air pollutants travel as part of an "airshed" to other states and provinces, and even to other countries or distant continents. The actual distance from the pollutant's source to the ultimate "sink," where it becomes a problem, may be short or very great, as much as several thousand miles in some cases.

Smoke from wood fires is an example of local air pollution that is as problematic today as it was for prehistoric peoples. The dependence of modern societies on oil, gas, and coal has contributed to local issues, including urban smog. Our reliance on these fossil fuels is also being felt on a much broader scale, with problems including acid rain and greenhouse gas emissions.

The impact of air pollution eventually affects all watersheds. Pollutants return to the earth's surface through precipitation and deposition; or they may affect ecosystems and people through changes in global climate. Once back at ground level, pollutants can immediately begin to harm living organisms as they travel downstream through the entire watershed. In this way, many air pollutants ultimately become a form of water pollution.

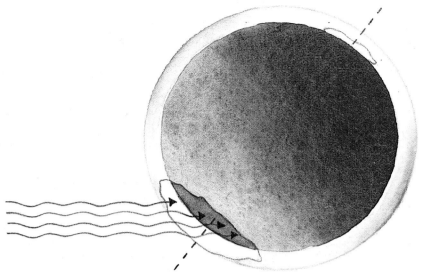

Reduced upper atmosphere ozone allows more ultraviolet radiation from the sun to reach the earth's surface.

## OZONE

Ozone ($O_3$) pollution appears as a frequent topic in environmental news stories. It is a confusing issue to understand, since ozone is at the same time both beneficial and harmful to life. From 6 to 25 miles (10–40 km) above the earth's surface, a layer of ozone shields us from harmful ultraviolet (UV) radiation, protecting us from skin cancer and other serious ailments. However, at ground level, ozone and other chemicals react and form photochemical smog, which irritates the eyes and affects the respiratory systems of many animals. Ground-level ozone is also harmful to plants, with estimated crop damage in the billions of dollars annually in the United States alone.

Humans are decreasing ozone levels high up in the earth's atmosphere, while also causing increased concentrations at ground level. This paradox involves a number of chemical equations. In the upper atmosphere, the main problem seems to stem from chlorofluorocarbons (CFCs), which are used in coolant for refrigerators and air conditioners, to make foam insulation and as a cleaning solvent. CFCs chemically break apart ozone molecules, leading to a decrease in the atmospheric ozone layer.

Down at ground level, pollution from automobiles and industry causes a separate set of chemical reactions. Nitrogen oxides, oxygen, and other pollutants react with sunlight to create ozone and even more pollutants, which we call smog. Smog is heavy, so it tends to stay put near ground level, making life miserable for most forms of life, especially during the hot, humid summer months.

## Good Ozone: The Ozone Layer

- Located high above the earth's surface, this natural protective layer of gases shields us from the sun's harmful ultraviolet rays. More UV light reaches the ground when this layer is thin.
- A huge hole in the ozone layer over Antarctica was discovered in the mid-1970s and was confirmed and widely reported about ten years later. This seasonal phenomenon has resulted in the loss of over 50 percent (and in some extreme cases as high as 95 percent) of Antarctica's natural ozone cover, and is worst between September and November. A similar hole in the ozone layer has also been discovered in the Arctic.
- Pollutants are causing a thinning of the ozone layer all around the globe, not just at the North and South Poles.
- Concern about the thinning ozone layer and possible human health effects has resulted in daily UV index readings in many communities.

## Bad Ozone: Problems at Ground Level

- Ground-level ozone, together with other pollutants from industry and vehicles, forms smog.
- Smog is harmful to the health of humans and other animals, as well as plants.
- Because it is heavier than clean air, smog remains close to the ground. In summer, hot, humid and polluted air often chokes large cities. This is especially true if they are located in valleys or other locations where there is limited air movement.
- Unfortunately, ground-level ozone cannot float upward to fill in the holes in the ozone layer in the upper atmosphere.

# THE GREENHOUSE EFFECT

Carbon dioxide ($CO_2$), moisture, and some other gases in the atmosphere help the earth to conserve or trap heat from the sun. Therefore, the surface of the earth stays warm in much the same way that the interior of a greenhouse is kept warm by its glass windows. Unfortunately, human activities, such as the burning of large quantities of fossil fuels and wood, are releasing ever-increasing amounts of carbon dioxide into the atmosphere. This means that more and more heat is trapped in the atmosphere, leading to a phenomenon called the greenhouse effect. Extensive logging around the world removes trees and other vegetation, which is unfortunate since plants absorb $CO_2$ from the atmosphere and help to reduce the problem.

One of the most likely results of increased greenhouse gas emissions is climate change, including global warming. The term "climate change" is perhaps the better choice of words, since not all areas of the world will necessarily get hotter. Climate change could trigger increased drought in many regions, more extreme weather variability, or even the melting of polar ice, with consequent higher sea levels and possible flooding in coastal regions.

Greenhouse gas emissions have been the subject of much debate recently, and the governments of many nations are scrambling to set targets for $CO_2$ reductions. The fact that greenhouse gases are increasing in the atmosphere is not the only issue. The more hotly debated questions relate to the specific effects, such as how much and how quickly temperatures will change, and how that will affect the climate and sea level in different areas. Regardless of one's predictions about the possible effects of climate change, it makes good ecological sense to reduce $CO_2$ emissions. Reducing greenhouse gas emissions goes hand in hand with reducing other air pollutants and efforts to protect and restore natural forest cover.

Ozone depletion and the greenhouse effect are environmental stories that are being widely covered by the media. In much the same way, acid rain has received a lot of press. But environmental problems do not come and go with media attention. All three issues remain important current concerns.

Incoming short-wave radiation

Greenhouse gas layer

Re-radiated long-wave radiation

• Greenhouse gases include carbon dioxide, methane, nitrous oxides, and CFCs. Carbon dioxide accounts for more than 50 percent of human-generated greenhouse emissions.

• When wood or fossil fuels are burned, the stored carbon they contain is released into the atmosphere in the form of carbon dioxide gas.

• Plants can absorb and use huge amounts of carbon dioxide when they grow and photosynthesize. This reduces greenhouse gases, and turns carbon back into a solid form that can be stored by vegetation.

• Incoming short-wave radiation from the sun passes through greenhouse gases, but much of the long-wave radiation which bounces off the earth's surface is trapped by greenhouse gases in the earth's atmosphere. As greenhouse gases increase, more of this radiation gets trapped.

• Greenhouse gases and the greenhouse effect occur naturally, but humans have dramatically increased the volume of these gases to the point where the world climate is being affected.

• The ozone layer occurs higher in the earth's atmosphere than the greenhouse gas layer.

# ACID RAIN

Acid rain is a term used to describe any form of precipitation that is more acidic than natural rainwater. The major sources of acid rain pollutants are from the burning of fossil fuels and the smelting of metals. These air pollutants react chemically with water vapor in the atmosphere and create acidic precipitation.

When we burn fossil fuels to produce electricity or to run our cars and factories, various waste products are released into the atmosphere. For example, gas or coal, which were once the carbon skeleton of prehistoric plants or animals, are transformed and liberated as carbon dioxide gas. Nitrogen and sulfur wastes are also produced, and react in the atmosphere to become nitrogen oxides and sulfur dioxide. These in turn combine with moisture in the air to form acidic solutions, which ultimately fall back down to earth as acid rain, snow, or sleet.

The ecological effects of acidic precipitation are numerous. In extremely polluted areas, lakes and rivers may become so acidic that adult fish are killed outright. Frequently, though, the effects are less obvious. Sometimes, adults can survive but their eggs and young cannot. The acid precipitation can cause soils to release aluminum and other pollutants, which are harmful to life. Invertebrates, such as clams, snails, crayfish, and aquatic insects, disappear as acidity increases, leaving little food for organisms higher up the food chain.

Acid rain also affects terrestrial habitats. Acidic water in snowmelt pools harms the eggs and young of frogs, toads, and salamanders. Acid rain alters the chemistry of soils and decreases the amount of many essential nutrients and minerals available, which has a detrimental effect on many plant species. Trees in the direct path of highly polluted air masses can show signs of decline – ranging from slowed growth and sickly boughs to death in extreme cases. Frequently, however, the effects are indirect. Acid rain may weaken trees, making them more susceptible to disease, drought, and other natural disturbances.

On a more positive note, many regions have soils and bedrock which are naturally able to neutralize or "buffer" acidic precipitation, thus minimizing the ecological consequences. In addition, technological advances continue to be made which significantly reduce the amount of pollution emissions which cause acid rain. Acid rain does, however, remain a serious environmental problem, and it appears that the road to recovery will be long.

## Acid Shock in Meltwater Pools

- Acid snow can accumulate over the winter months in northern regions. When it melts in spring, an entire season's acid content is released.
- This large pulse of pollution creates an acid shock in small streams and pools.
- Some species, such as amphibians, breed in these meltwaters, and the acid shock kills developing eggs and harms the larvae. Salamanders, toads, and frogs are sensitive to many pollutants and to habitat destruction. These environmental monitors have disappeared from many regions of North America.

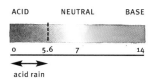

### pH Scale

- The pH scale is confusing – when acidity goes up, the number on the pH scale goes down. And there is a tenfold increase in acidity for each one unit down on the logarithmic pH scale.
- When rain is pH 4.6, it is actually ten times more acidic than clean rainwater, which is pH 5.6; rain which is pH 3.6 is actually a hundred times as acidic as unpolluted rain.
- Clean rainwater is naturally a little bit acidic (pH 5.6) because carbon dioxide and moisture combine to form a weak carbonic acid solution.
- Acid returns to earth in many forms – rain, snow, fog, sleet, hail, and even dry deposition.

## Dead Lakes

- Dead lakes are deceptively beautiful. The increased clarity of the water and the aquamarine blue color are associated with increased acidity. Don't be fooled; this is a very bad sign.
- As the pH drops, biodiversity decreases, gradually leaving fewer and fewer species.
- A moderately acidified lake may contain some large fish, but few of their offspring will survive. Invertebrates, including crayfish, aquatic insects, and clams, do not tolerate acidic water, which obviously affects the food supply for larger creatures.

## Forest Decline

- Trees located very close to a major point source of pollution can be killed.
- Trees farther from the source of the pollutants grow more slowly and have sickly or dying branches.
- Forests growing in very foggy climates are greatly affected because the trees are frequently bathed in moist, acidic air.

## Effects of Soils and Winds

- The biggest air pollution problems occur downwind from major industries and cities.
- Unfortunately, some of the areas most sensitive to acid rain are also located downwind, such as northern Ontario and Quebec and the Adirondack Mountains of New York.
- Many regions have soils and rocks that act as natural buffers against acid rain. If you live in an area with lots of limestone, acid rain is probably not a big problem in your large lakes and rivers.

## AIR POLLUTION

There are a great many simple things that individuals can do to help reduce air pollution. These actions help to keep the air cleaner and the atmosphere healthier. And because many pollutants end up falling back down to the earth's surface, cleaner air often means fewer problems on land and less water pollution. Clean air and healthy watersheds go hand in hand.

**Pollution from automobiles contributes to ground-level ozone, greenhouse gases, and acid rain, so why not consider:**

* cycling and walking more often – and driving less (especially during hot, smoggy summer weather)
* carpooling
* taking public transportation in the city
* combining several errands into one trip if you must use the car
* using higher octane and cleaner gasoline, which are more efficient and pollute less
* replacing older vehicles with fuel-efficient or alternative-fuel vehicles
* keeping vehicles well maintained to minimize pollution, and checking tire pressure and wheel alignment
* driving at a lower speed to reduce gas consumption and pollution
* avoiding "jack-rabbit" starts and idling cars for more than a couple of minutes

**In our homes, we can also help to reduce air pollution by:**

- making sure our houses are well insulated against cold and heat
- keeping our thermostats a little lower to save fuel (and money)
- heating with high-efficiency furnaces, using the cleanest fuels available
- purchasing refrigerators and air conditioners with environmentally friendly coolants
- avoiding products that contain environmentally harmful components, such as CFCS
- planting trees and shrubs, which helps to reduce the amount of carbon dioxide and other pollutants in the atmosphere
- avoiding the use of hazardous household products that pollute the air when burned after consumer use
- disposing of all hazardous household products (for example, used paints and motor oils, old tires) in special hazardous waste collections rather than with regular garbage
- raking leaves by hand and avoiding the use of noisy and polluting leaf-blowers
- avoiding the use of snow-blowers and instead shoveling manually, if possible

**At the cottage and on the farm, we can help to reduce air pollution by:**

- enjoying and exploring the outdoors through non-polluting activities such as hiking, canoeing, kayaking, sailing, and cross-country skiing
- reducing or eliminating our use of motorized outdoor vehicles such as jetskis, snowmobiles, all-terrain vehicles and power boats
- replacing old boat engines with newer, quieter ones that use much less gasoline and produce less air and water pollution
- composting leaves and twigs or leaving them to decompose in the woods, rather than burning them

**We can also pressure politicians and industry to reduce air pollution by:**

- encouraging cleaner industrial technologies and legislating mandatory environmental standards
- installing pollution control scrubbers for coal-burning utilities and metal smelters
- replacing antiquated factories with cleaner, more efficient ones
- switching to alternative sources and cleaner fuels to generate electrical power

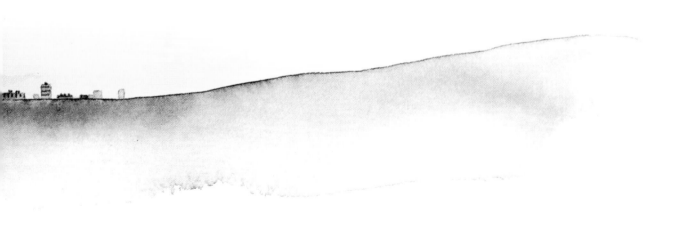

# 6

# WATER
# POLLUTION

Water pollution occurs in many different forms and in all parts of the world. Some poisons and toxins are readily identifiable as pollution, and their environmental effects are easy to pinpoint. In other cases, the problem is harder to see and more difficult to understand. For example, a cyanide chemical spill will affect the watershed immediately by killing fish and other aquatic life. Nitrogen and phosphorus pollution, on the other hand, is much more complicated. These essential nutrients are needed by plants and animals, but if too much enters our waterways from farms and cities, they become a serious form of water pollution.

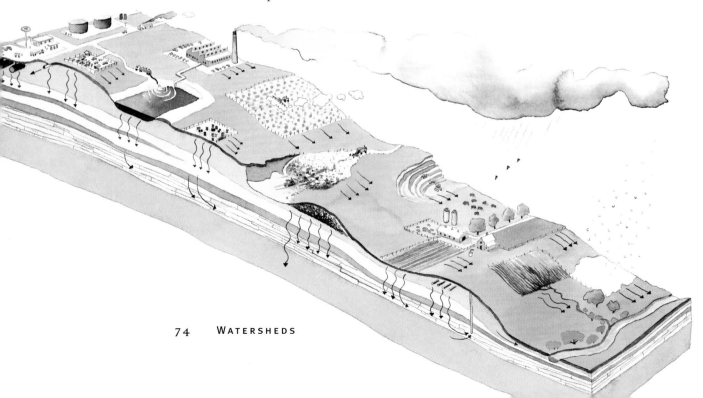

Sometimes, the source of the problem is easy to spot, as is the case with major "point sources" of pollution like individual factories or mines. The challenge here is to encourage, legislate, and enforce pollution reduction measures. Pollution, however, comes from not only these sorts of industrial sources, but also from innumerable "non-point sources," including millions of houses, apartments, cottages, and farms. In many respects, pollution from non-point sources is more difficult to locate and control because it involves so many people, and comes from so many different sources. In order to successfully improve air, soil, and water quality, all sources of pollution must be targeted, including industry, households, farms, and the day-to-day actions of each and every one of us.

# Bioconcentration and the Effects of Contaminants

One of the most horrific examples of water pollution can be illustrated by examining the environmental impact of certain long-lived chemicals, such as the organochlorine contaminants and heavy metals. Toxic chemicals enter aquatic food chains through air pollution as well as direct discharge into water courses. While the concentration of toxins is usually low in water samples, it is often very high in fish and fish-eating animals. The following account demonstrates how even minute quantities of these pollutants can become incorporated into the tissues of living creatures and become increasingly concentrated higher up the food chain.

Living organisms continuously take nutrients and food from their surroundings. At the same time, they also accumulate pollutants present in the water through a process called bioaccumulation. Regardless of where an organism is on the food chain, it bioaccumulates pollutants throughout its life. While some pollutants may be released back into the water or passed on to offspring, much of it remains within the body of the organism.

The concentration of toxic pollutants becomes even more amplified higher up the food chain. At the base of the food chain, microscopic aquatic plants gather nutrients and bioaccumulate pollutants in the water. Small animals, such as zooplankton, feed on large quantities of these phytoplankton and therefore accumulate even higher concentrations of chemicals. In turn, vast numbers of small fishes are eaten by the larger fishes. Many of these fish are then eaten by eagles, osprey, herons, otters, seals, whales . . . and people. With each step up the food chain, the concentration of pollutants increases. This is called biomagnification. The concentration of pollutants in animals at the top of the food chain can be hundreds of thousands – even tens of millions – times greater than in the surrounding waters.

Among the longest lasting and most harmful pollutants are the chlorine-based organic compounds (or organochlorine contaminants), which become stored in the fatty tissues of living things. Many industries have made products or wastes that contain polychlorinated biphenyls (PCBs), or make other toxic substances that contain dioxins and furans. DDT is another example of a highly persistent toxin, which was used extensively as a pesticide in the twentieth century. While DDT has been banned in Canada and the United States since the early 1970s, it is still used in other parts of the world. Despite its restricted use in North America, tissue samples from many animals still contain residues of DDT or its resulting chemical by-products. Heavy metals, such as lead, mercury, and copper, are another form of bioaccumulating pollution that is very harmful ecologically. They are stored in tissues such as muscle and even the brain.

The effects of DDT, PCBs, and other toxins on wildlife can be horrific. During the 1950s and 1960s, the populations of many predatory bird species crashed. Fish-eaters, such as bald eagles, osprey, herons, and cormorants were affected, as were other carnivorous species, including the peregrine falcon. The result was reproductive failure: eggs were often infertile, or had such thin shells that the weight of the parent crushed them. In some cases, young hatched but soon after died because of their deformed, twisted beaks. Other animals, such as some fish species, developed strange tumors and other diseases.

Fortunately, conservation efforts and tighter regulations, together with increased public awareness and greater industrial accountability, have helped to decrease the concentrations of many pollutants, including PCBs and DDT. Positive trends are emerging throughout North America, and some predatory birds are making a comeback. Unfortunately, as mentioned above, many extremely dangerous pollutants are still being widely used in other parts of the world. This is harmful for local wildlife, for migratory species, and for ecosystems as far away as the Arctic, which can receive pollution from atmospheric fallout and through ocean currents.

The ecological effects of pollutants are very difficult to assess. For example, some marine mammals have accumulated large amounts of contaminants in

• Bioaccumulation is the increase in pollutants in an organism over its lifetime. Each time an animal eats or drinks, it is also consuming pollutants, much of which remains within its body.

• Some species, such as clams and mussels, filter water in order to capture and eat microscopic plankton. Because they have to filter large quantities of water in order to get enough food to eat, they also bioaccumulate a lot of pollutants. Many areas downstream from heavy industries have posted signs warning people not to eat the contaminated clams and mussels.

their stores of fat-rich blubber. Many scientists believe that this is reducing their reproductive success and causing increased frequency of serious diseases, such as cancers. While most people believe that pollutants continue to harm wildlife, it is often very difficult to prove conclusively that a particular industry or chemical is directly responsible. The number of factors affecting ecological food webs is enormous, so clear proof in the form of a "smoking gun" remains elusive.

## BIOMAGNIFICATION

**CORMORANT**

(tertiary or higher order consumer –
500,000 to 50,000,000 times pollution concentration of water)

**SMALL FISH**

(secondary consumer –
100,000 to 500,000 times pollution concentration of water)

**ZOOPLANKTON / AQUATIC INVERTEBRATES**

(primary consumer –
10,000 to 50,000 times pollution concentration of water)

**PHYTOPLANKTON**

(producer –
1,000 to 5,000 times pollution concentration of water)

● Plants and animals bioaccumulate pollutants during their lifetime. In addition, larger organisms eat quantities of smaller prey, which increases or biomagnifies the amount of pollution with each step up the food chain.
● Toxins may be discharged directly into the water, or they may get washed in from the land. Eventually they accumulate in larger rivers and lakes. Therefore, predatory species of animals which eat large quantities of fish or other aquatic animals tend to be the most seriously affected.
● The concentration of pollution in large fish or birds can be hundreds or thousands of times higher than in plankton, and millions of times more concentrated than in the water itself.

# INDUSTRIAL SOURCES OF WATER POLLUTION

There are many different sources of water pollution, and industry is frequently blamed as one of the worst offenders for environmental damage. Large factories with billowing smokestacks and offensive water outflows are easily targeted as major point sources of pollution. In addition, many polluting industries are also associated with the loss of large areas of natural habitat, such as in the case of forestry or strip mining.

Factories are the source of various forms of pollution. Smokestack emissions cause air pollution, waste waters can add pollutants directly to rivers and lakes, and groundwater moving through soils transports toxins from one region to another, where it may enter rivers and lakes. Air pollution, such as acid rain, eventually falls back down to earth and becomes a form of water pollution far from its original source. And industrial accidents, such as the spillage of toxins, results in the contamination of soil, aquifers, and lake and river water.

Some of the worst industrial pollutants to affect water bodies directly are the highly persistent chlorine-based contaminants, heavy metals, and toxic hydrocarbons formed from oil and gas. Accidental oil spills can have far-reaching ecological effects. Factories that use water to cool machinery produce warm water, which when discharged into waterways can affect wildlife – a form of thermal pollution. Even if the outflow is clean, the increased temperature reduces the amount of oxygen in the water and dramatically alters local habitats.

The amount of pollution being produced, together with the huge variety of chemical contaminants, creates a depressing picture of disease and death. Advances continue to be made, though, in cleaner, greener technology which can be used to reduce the amount of pollution. New, more environmentally friendly factories are being built, and some older industries are being reconditioned to reduce pollution.

## Oil Spillage

- Oil pollution can come from many sources. Some spills are extremely large, as in the case of a major oil tanker accident at sea. Thousands of small oil spills probably occur every day, both on land and on the water. Many of the effects are similar, though, regardless of whether the source is from ship bilge cleaning, leaky storage tanks, or someone pouring dirty automobile oil on the ground or down a sewer.

- Oil refineries and transfer stations are other potential sites for oil contamination.

- When oil is spilled on lakes and the ocean, seabirds are among the most drastically affected. Oil-soaked feathers do not provide good insulation or flotation, so seabirds often die of exposure to cold water or they may drown.

- Seabirds congregate in large groups, especially near important feeding or nesting grounds, which makes them highly susceptible to spills. Hundreds of thousands of seabirds have been killed during large oil spills at sea.

- Seabirds routinely wash up dead and oiled on beaches, even when no major spills have been recorded. These mortalities may be caused by small oil slicks from bilge cleaning or thoughtless handling of waste oil.

- After a spill, oil spreads across the water's surface; in addition, some crude oil evaporates and some becomes mixed into the water below.

- Direct contact with spilled oil can sicken and kill birds, mammals, fish, aquatic plants and seaweeds, as well as invertebrates, such as lobsters, anemones, barnacles, and sea stars.

- Even low concentrations of oil particles in water can become assimilated into plant and algae tissues. Some of this vegetation is eaten, and pollutants then get passed up the food chain, becoming increasingly concentrated with each step.

- Fortunately, natural biological processes help to clean up spilled oil products, and most shorelines are relatively clean and healthy within a few years. However, wildlife species with low reproductive rates, such as seabirds, may take decades to recover their populations.

- Oil and its by-products are toxic, and even lethal, to plants and wildlife.

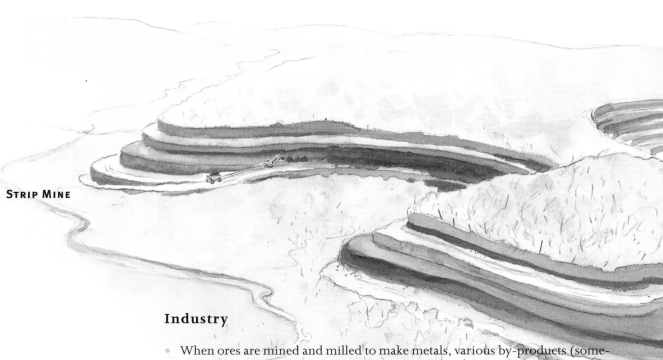

**STRIP MINE**

## Industry

● When analyzing reports about pollution, it is important to note that not all pollutants have the same potency. Dioxins and PCBs are both chlorine-containing chemicals, but dioxin is much more dangerous – even at significantly lower concentrations.

● When ores are mined and milled to make metals, various by-products (sometimes called mine tailings) are produced, including lead, nickel, arsenic, cadmium and other elements. The runoff from strip mines is often extremely acidic. Smelters which then refine the metal products release more pollutants.

● Heavy metals are substances, such as mercury and lead, that are highly toxic to wildlife and humans, even at very low levels. They become more concentrated in species at the top of food chains.

● PCBS (polychlorinated biphenyls) are no longer being produced in North America. They were used widely for many years in electrical transformers, fluorescent lights, and for numerous other household and industrial purposes. Disposal of PCBS and the cleanup of PCB-contaminated land is a major environmental chore. PCBS remain potent for many years, and wildlife in many regions is still highly contaminated.

● Pulp and paper mills produce both air and water pollutants. Chlorine, used in the bleaching process, forms numerous hazardous compounds, some of which are very long-lived in the environment. Dioxins and furans, for example, are exceedingly toxic by-products of pulp and paper processing.

● The forestry industry also contributes to water pollution when forests are sprayed with pesticides. Although the use of the highly toxic and persistent pesticide DDT has been banned in Canada and the United States since the early 1970s (with a few exceptions), it is still widely used elsewhere in the world. Other pesticides may be shorter-lived and less harmful, but they all have some environmental costs.

## Thermal Pollution

- Many industries use enormous quantities of water to cool their machinery. When this warm water is released into local rivers and lakes, it changes the aquatic ecosystem and can affect the creatures living there. Among other effects, warm water holds less oxygen than cool water, which can have a negative impact on many species of fish and invertebrates.

## Radioactive Wastes

- Nuclear power plants can produce electricity without releasing acid rain pollutants, and without interrupting the flow of rivers. They do, however, produce radioactive wastes which are dangerous for an extremely long time (from decades to millennia). In addition, they generate a lot of thermal pollution in the form of heated waste water.

**NUCLEAR POWER PLANT**

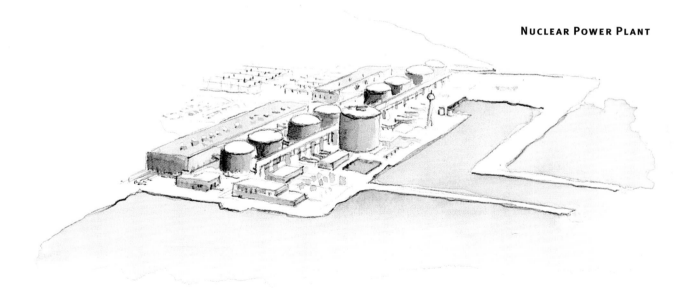

# Urban Sources Of Water Pollution

Urban areas are complex human spaces, densely packed with homes, businesses, and industries. To service these built structures, we set aside enormous amounts of land for the supporting infrastructure, such as roads, parking lots, pipelines, and electrical power corridors. One way or another, the human activities associated with these intrusions contribute to air and water pollution.

Many of us forget that water runoff from rooftops, roads, and lawns eventually ends up flowing into urban streams, rivers, and lakes. The same thing happens with water that is flushed down sinks and toilets. Domestic waste water can include hazardous household materials which we use for cleaning and other chores. Excess nutrients from sewage and fertilizers pose one type of water pollution problem, and sewage, which contains bacteria, can cause serious diseases. Sewage treatment varies considerably from place to place; a great many cities still discharge untreated sewage directly into waterways, and others have only basic treatment plants. Chlorine is used effectively to kill germs in homes and in sewage treatment and water filtration plants – but once chlorine enters streams and lakes, it can threaten wildlife.

Other urban sources of water pollution include oil and salt that get washed from roads, liquid wastes that leak from massive landfill sites for garbage, pesticide, and herbicide residues, and the toxic chemicals that some people still pour down sinks. We should all try to minimize our use of toxic chemicals and do all we can to keep our waters clean. After all, whether it comes from a well underground or from a nearby lake, we all end up drinking the same water.

## The Moral of Summer Thunderstorms

Heavy thundershowers are common in many regions during the hot and humid days of summer. During these storms, large amounts of rain falls, and in most cities this water quickly flows from rooftops, roads and driveways into sewers.

This is a bad time of year for watersheds because large amounts of pollution get washed into rivers and lakes along with the storm water. Heavy rains wash fertilizer and pesticides from lawns and gardens, and the extra volume of water can combine with waste water from sinks and toilets to overload sewers and sewage treatment plants. Excess nutrients and bacteria from raw sewage end up polluting rivers and lakes, and many cities routinely close swimming beaches

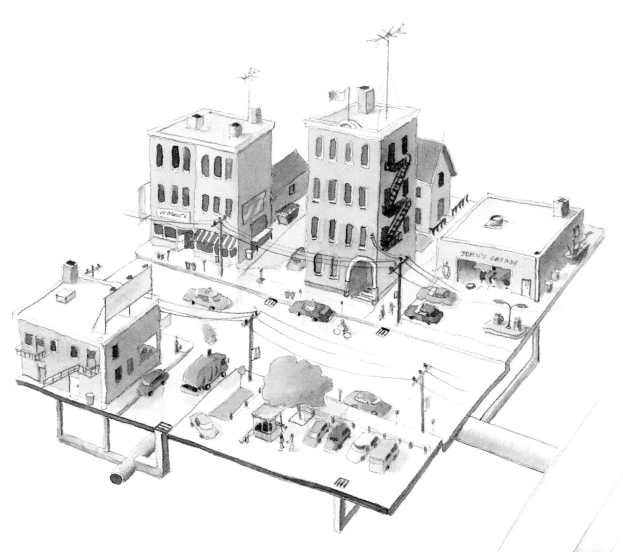

after summer storms as a health precaution. The large volume of runoff creates dangerous flash flood conditions in small creeks and large rivers alike; drownings are not uncommon.

These summer storms, although sometimes very violent, are usually short-lived, and the sun may be shining again within a few hours. Ironically, it is common to see people watering their lawns later the very same day. This does not make sense – ecologically or economically. We should act more sensibly and harness this water by putting it to use. Let's treat water as a valuable resource. This also makes sense financially, since many people are on a water metering system: the less we use, the smaller the water bill.

## Sewage Treatment and Water Filtration

*Where does it go when you flush the toilet?*
*Where does it come from when you turn on the tap?*

In cities, water which goes down drains passes through sewer systems and in many cases goes to a sewage treatment plant. Then, after treatment, it is discharged into lakes and rivers. In many cities, pipes withdraw drinking water from these same lakes and rivers and carry it to a filtration plant for treatment. Municipally treated water is then distributed to consumers through very extensive systems of pipes. Towns and cities that are not located near large lakes or rivers may get their water from underground aquifers.

The fact that we often get our drinking water from the same bodies of water into which our sewage goes is a rather sobering thought. This is even more disturbing when you consider the thousands or millions of people who live in a city, and the fact that treatment plants are not generally equipped to remove toxic contaminants from water. As you move farther downstream within each watershed, the cumulative effects of water pollution become greater and greater.

### Sewer Systems

● There are several different kinds of sewer systems. Cities and towns may use one type or another, or frequently a combination of systems.

● In the "combined sewer system," runoff from roofs, driveways, and roads combines with water from drains and toilets. Under optimum conditions, all of this goes to a sewage treatment plant.

● During heavy rains, the extra volume from roads and rooftops can overload the system. This "combined sewer overflow" (or CSO), which includes raw sewage, may then be released directly into watercourses.

● Some cities are building massive storage tanks and tunnels to hold the overflow until treatment plants can process it.

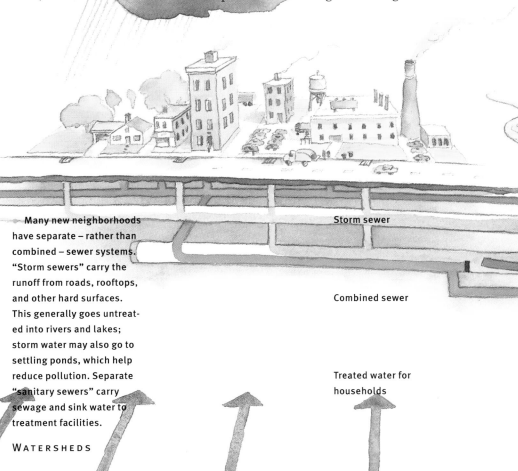

● Many new neighborhoods have separate – rather than combined – sewer systems. "Storm sewers" carry the runoff from roads, rooftops, and other hard surfaces. This generally goes untreated into rivers and lakes; storm water may also go to settling ponds, which help reduce pollution. Separate "sanitary sewers" carry sewage and sink water to treatment facilities.

Storm sewer

Combined sewer

Treated water for households

## Sewage Treatment

• During primary treatment, solid objects are removed from sewage by screens or filters, and suspended solids are settled out in large tanks.

• Secondary or "biological" treatment reduces the amount of biodegradable organic wastes in sewage. Bacteria and other microorganisms chemically break down the sewage; this process consumes large amounts of oxygen.

• Secondary treatment is an important step because it reduces the "biochemical oxygen demand" (or BOD). Biochemical oxygen demand indicates the amount of oxygen needed by nature to break down sewage and organic materials. Sewage discharge with a high BOD will deplete the amount of oxygen in the water body that receives the effluent.

• Secondary treatment can be accomplished using sewage lagoons or large tanks with "trickling filters." The "activated sludge process" mixes and aerates sewage to ensure microorganisms decompose wastes. A settling pond then separates liquid and solid components.

• After treatment, most sewage plants add chlorine to kill bacteria, then discharge water to lakes, rivers, or the ocean.

• The disposal of sewage sludge from primary and secondary treatment remains a problem. It is either sent to landfills, burned, dumped at sea, or sometimes used as fertilizer.

• Some modern plants add an additional "tertiary" treatment step to further reduce BOD, since secondary treatment removes only about 50 percent of nitrogen and 30 percent of phosphorus. Most facilities do not remove toxic substances.

• A large percentage of cities and towns still discharge untreated sewage directly into rivers, lakes, or the ocean. Improper or incomplete sewage treatment remains one of the biggest sources of water pollution.

Sewage treatment plant

Sanitary sewer

Combined sewer overflow

## Water Filtration

• Most municipalities have water filtration plants to clean and purify water.

• Large particles are removed from untreated source waters, and alum is added to settle out smaller particles.

• Water then passes through a sand filter to remove particles and some bacteria.

• Chlorine is added to kill bacteria and other microorganisms. Fluoride may be added to reduce the incidence of dental cavities. Water then travels through a complex distribution and storage network on its way to houses, businesses, and industries.

Water filtration plant

## Alternative Sewage Treatment

One of the biggest problems with sewage and storm water runoff is that it contains a high concentration of decomposing organic matter. When bacteria and other microorganisms break down these wastes, they use up a lot of the oxygen in the water, which has a negative impact on many aquatic organisms.

Some small communities are creating experimental filtration systems and artificial wetlands and storm-water retention ponds to treat sewage and storm water. In these alternative processes, decomposition occurs in a series of marsh-like treatment cells, which may be either indoors or outdoors. This means that local waterways do not experience as much oxygen depletion because the outflow has a lower concentration of decomposing organic matter (and therefore lower BOD).

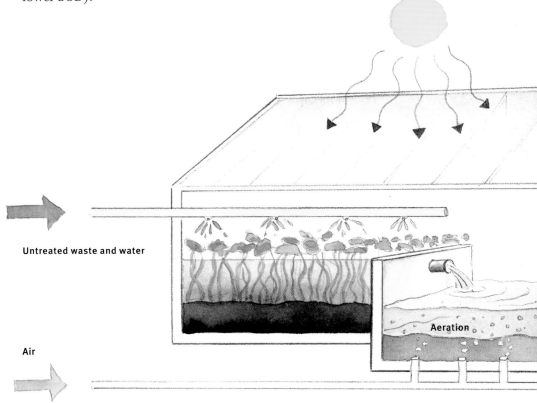

Untreated waste and water

Air

Aeration

Marsh plants, such as cattails, grow very quickly and are therefore extremely efficient in removing large amounts of phosphorus and nitrogen from untreated water. They are also very effective at soaking up heavy metal pollutants; the problem remains, though, of what to do with the plant tissues that have absorbed these toxic pollutants. In addition to breaking down wastes and absorbing excess nutrients, these artificial wetlands mimic natural processes and help to filter the water and remove sediments. By the time water exits the system, it is much cleaner and also much clearer.

Storm-water retention ponds are an increasingly common requirement in new residential and commercial developments. Careful study is required to ensure that the size of these ponds or facilities is large enough to support the volume of water in the so-called "sewershed." These ponds are designed to reduce the pressure on traditional sewage treatment plants, but they can also provide habitat for a diversity of wildlife.

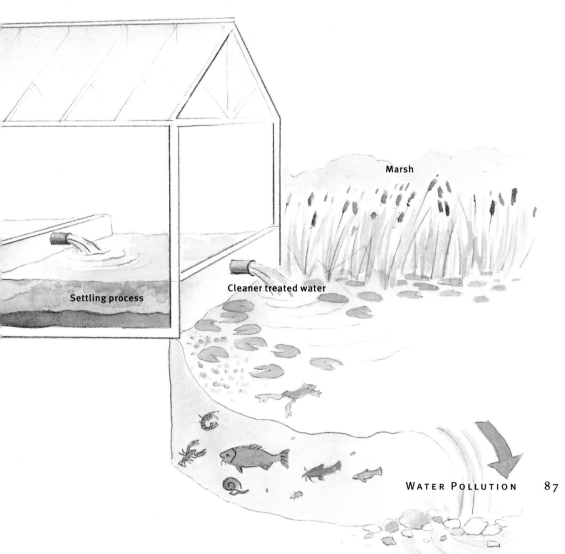

Marsh

Settling process

Cleaner treated water

# Nitrogen and Phosphorus Pollution: Too Much of a Good Thing?

One of the biggest water-quality problems in many regions is nitrogen and phosphorus pollution. In some ways, this issue is more difficult to understand than pesticide or heavy metal pollution. With those toxins, any amount is bad and higher concentrations are simply worse. Nitrogen and phosphorus pollution is different; these nutrients are required by both aquatic and terrestrial plants in order to grow and survive. In turn, all animals also need these nutrients and obtain them by eating plants or other animals. The central point with nitrogen and phosphorus, therefore, is the amount — too much, and they pollute waterways.

- Nitrogen and phosphorus enter waterways from many places. The biggest sources are sewage wastes, livestock manure, and fertilizers from farms, gardens, and lawns.
- These excess nutrients travel overland and through tributary streams; the water cycle carries these nutrients downstream to lakes. The larger the watershed, and the more farms and people within it, the bigger the problem.
- Nitrogen and phosphorus both contribute to this problem, but phosphorus is the greater concern since plants need very little for growth. With phosphorus, "a little goes a long way."

- The excess of nitrogen and phosphorus causes rapid growth of algae, cyanobacteria, and other floating aquatic plants. These "blooms" make the water extremely murky (or turbid). In extreme cases, you could be standing in ankle-deep water and be unable to see your toes.

- As the water becomes choked with this algal bloom, sunlight is unable to penetrate very deep into the water column. Plant material that is farther from the surface dies and falls to the bottom of the lake.

- The resulting large quantities of dead plants and animals on the lake bottom lead to oxygen depletion. Microorganisms that decompose this matter use large amounts of oxygen, eventually consuming most or all of the oxygen from the water at the bottom of the lake.

- Fish and other oxygen-dependent species that live near the bottom are then in peril. In some years, huge fish die-offs have occurred. As oxygen levels decrease, so does the biodiversity.

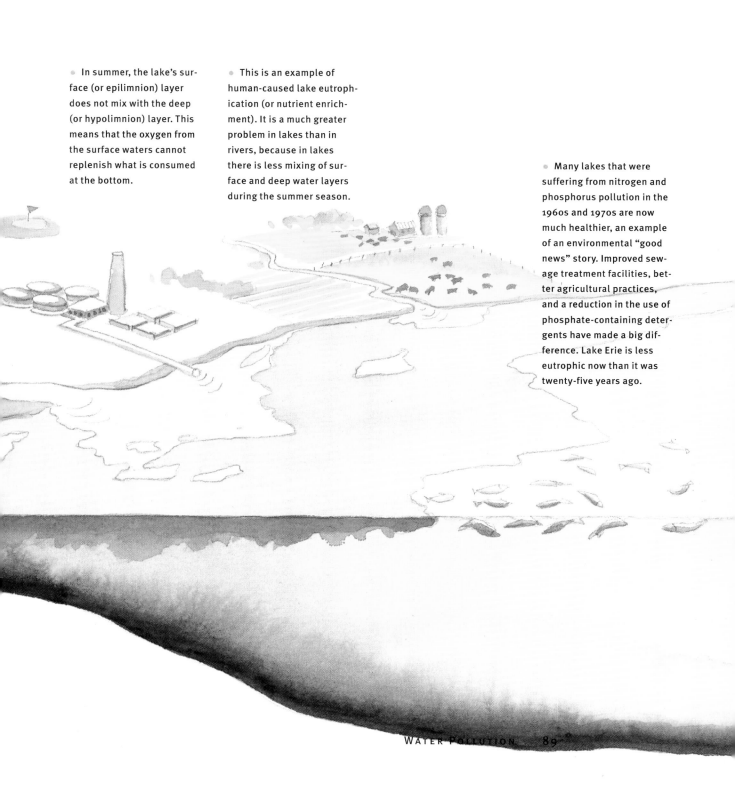

- In summer, the lake's surface (or epilimnion) layer does not mix with the deep (or hypolimnion) layer. This means that the oxygen from the surface waters cannot replenish what is consumed at the bottom.

- This is an example of human-caused lake eutrophication (or nutrient enrichment). It is a much greater problem in lakes than in rivers, because in lakes there is less mixing of surface and deep water layers during the summer season.

- Many lakes that were suffering from nitrogen and phosphorus pollution in the 1960s and 1970s are now much healthier, an example of an environmental "good news" story. Improved sewage treatment facilities, better agricultural practices, and a reduction in the use of phosphate-containing detergents have made a big difference. Lake Erie is less eutrophic now than it was twenty-five years ago.

# AGRICULTURAL AND RURAL SOURCES OF WATER POLLUTION

Many of the water pollution issues that trouble urban areas are equally problematic for rural areas. Take, for example, the question of pesticides, herbicides, fungicides, and all the other "-cides." Whether used in the city or the country, hazardous chemicals can leach through soil into groundwater or flow into streams and lakes, having ecological consequences in many places.

Fortunately, most of the products used today are chemically short-lived, meaning that they break down into less harmful substances fairly quickly. This makes them a great deal better than earlier pesticides, such as DDT, which remains toxic for about fifty years. Nonetheless, they are effective in killing certain pest plants and animals, and there are many unresolved questions about the more subtle, long-term health and environmental concerns for creatures great and small.

Chemical fertilizers and livestock manure contain the raw nutrients that plants need to grow. Therefore, they are often used to increase crop productivity and to keep soil from becoming infertile. While this is not a bad thing, it can cause major problems down the road. Some of the fertilizer and manure ends up washing into watercourses. This is particularly likely during heavy rainstorms and spring snowmelt, especially if the fields and livestock pastures are close to water.

There are tens of millions of farms and cottages in the large North American drainage systems. Together, the magnified effect of excess nutrients entering waterways can cause eutrophication and oxygen depletion in lakes and rivers. The rich topsoil itself can get washed away, increasing the need for more fertilizers and clogging up rivers. Fortunately, major improvements have occurred during the last ten to fifteen years: farming techniques and sewage treatment are more effective, and the amount of phosphorus used has decreased.

Cottages and farmhouses add to rural water pollution when sewage waste leaks from faulty septic systems. This should be a major health concern for the rural public, since most people in rural areas get their drinking water from wells. No one wants to contract diseases caused by bacteria from human or animal feces. Likewise, there is serious public concern about the uncharted long-term health risks associated with even very low amounts of chemical contaminants in drinking water.

## WATER POLLUTION

In our homes and farms and cottages, we can all do a large number of simple things to reduce water pollution. Whether properly maintaining your septic system or avoiding the use of toxic cleaning supplies or pesticides, positive actions can help to minimize negative environmental impacts. By putting pressure on industry and governments, we can let others know that clean water and environmental responsibility are critically important to all of us. While industries need to clean up their act, individuals like ourselves must also do all we can to help out.

Using less water also helps to reduce pollution. Water conservation decreases the amount of water being treated with chlorine in sewage and filtration plants. It also decreases the risk of sewer overflows, and leaves more water in rivers, streams, and underground aquifers, which is good for wildlife. It is also good for us, since many regions experience water shortages during parts of the year. Conserving water will probably save a lot of money, too, on your water and sewage bills.

**Porous driveway**

**Curb and gutter**

**Around Home:**
**Helping Outdoors**

• Road salt is harmful to many plants and animals. Shovel your walk and driveway and use sand, cat litter, or calcium chloride instead of salt whenever possible.

• Water your lawn in the early morning and when there is no wind. When it's hot or windy, most of the water evaporates, leaving you with a big water bill and thirsty grass.

• Use a bucket and sponge when washing your car, then rinse off quickly with a hose.

• Sweep walks and driveways instead of hosing them down to remove grass clippings and leaves.

• To keep your grass healthy and reduce the need for watering, keep mower blades very sharp and set for a cutting height of about 2–2½ inches (5–6 cm). Select a type of grass that suits your soil and climate. Keeping grass healthy is also the best way to avoid insects, pests and weeds.

• Leave grass clippings on your lawn as free fertilizer. Leaves are free fertilizer, too, and great for your garden. They provide essential nutrients, such as carbon and nitrogen, and shade the soil, reducing the need for watering.

• Pull weeds by hand or learn to live with them. Check with your local garden supplier about organic alternatives to pesticides and chemical fertilizers.

• Instead of grass, use native ground cover plants or "naturalize" your yard. You'll save time and energy and provide a home for native wildlife.

• Disconnect your downspouts from sewer systems and collect the rainwater in barrels. Connect a hose and use this "free" water in your garden.

• Reducing the amount of water going into sewers is beneficial. For example, gravel driveways mean that rain percolates through the soil and is naturally filtered, rather than entering sewers. Rain barrels and other water conservation measures all help.

• In the house and garden, make sure to dispose of hazardous materials safely. Most towns and cities now have safe disposal centers or door-to-door pickups. Better yet, try the alternative cleaners listed on page 97.

Rain barrel

Sanitary sewer

Storm sewer

Avoid using chemical fertilizers and pesticides.

## Around Home: Helping Indoors

- Install water-saving showerheads, take shorter showers, and use less bath-water.
- Install a low-flow toilet. Or, install a toilet dam for your current one which will reduce the amount of water used for each flush. You can make your own toilet dam by putting a water-filled plastic container in the tank. (Make sure it does not interfere with moving parts.)
- Turn off the tap while brushing teeth or shaving.
- Install aerators on all taps; these reduce the amount of water used but keep the pressure high.
- Repair leaking taps right away. Also, check to see if your toilet leaks by adding a few drops of food coloring to the tank – if water is leaking, there will be traces of color in the bowl. It is easy to replace the defective parts.
- Keep a pitcher of water in the refrigerator, rather than running the tap for a long time to get cold water. You can also wash fruits and veg-etables in a partially filled sink or container, then rinse.
- Use dishwashers and washing machines only when you have a full load. Don't use garburetors, which waste water and increase BOD in lakes and rivers.
- Start composting your vegetable wastes to reduce your garbage by more than one quarter. It makes great fertilizer for the garden!
- Make sure hazardous household wastes do not end up in the garbage, as they can leach into soil and groundwater from landfill sites; save for hazardous waste disposal.
- Avoid purchasing toxic products; use alternatives instead. For example, clean windows with a diluted vine-gar solution and wipe with old newspapers. See page 97 for alternatives to toxic cleaners.

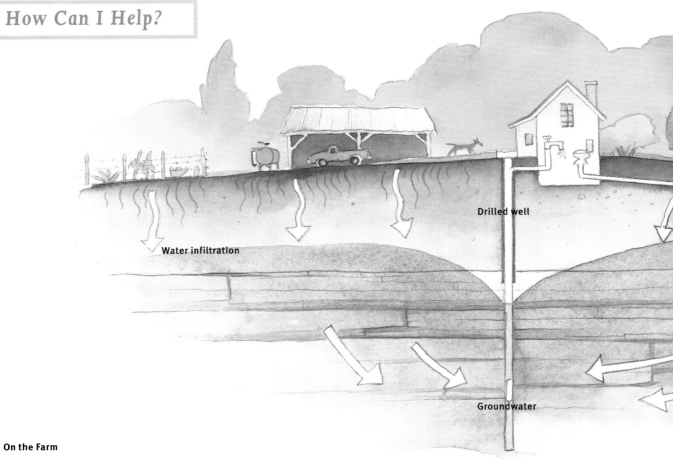

Drilled well

Water infiltration

Groundwater

### On the Farm

- Keep livestock fenced away from rivers and lakes to reduce soil erosion and pollution from manure and sediments.
- Manure can be safely stockpiled in covered, concrete storage areas and used as a source of fertilizer for fields. Solid manure should be spread on dry ground and tilled within one day. Liquid manure should be injected directly into the soil.
- Manure should never be spread on frozen or very wet fields or when rain is likely, since much of it will wash away. This pollutes waterways and wastes an important source of fertilizer.
- Rotate crops and include fallow periods so that nitrogen can be replenished naturally in the soil by decomposition and nitrogen fixation.
- Leave permanent natural buffers of trees and shrubs along ditches, streams, ponds, and lakes. This will reduce soil erosion, shade the water and keep it cool for fish and fishing, and provide habitat.
- Leave plant residues (i.e., rooted plant stalks) from the previous year's crop to reduce topsoil erosion. The loss of nutrient- and mineral-rich topsoil remains one of the biggest environmental problems facing agricultural areas.
- Plow along the contours (rather than straight up and down the slopes) to reduce erosion and the need for some fertilizers.
- Plant trees as windbreaks to help reduce soil erosion by wind. Grassy access paths to fields are better than exposed soil.
- Use fertilizers only when essential. Eliminate or minimize use of chemical pesticides and research the least harmful products.
- Irrigate crops in the early morning on calm days to minimize water loss through evaporation.

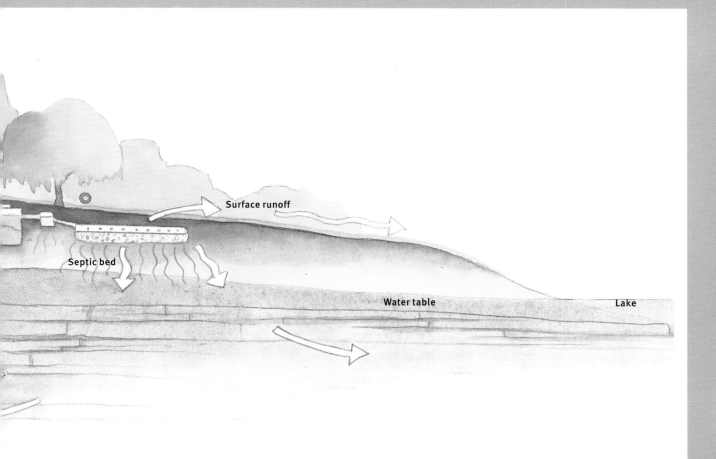

Surface runoff

Septic bed

Water table

Lake

## Rural Homes and Cottages

- Most rural houses and cottages have septic systems to treat sewage, which can be effective in removing excess nutrients and reducing bacteria. They must be properly installed, maintained, and occasionally pumped out to maintain maximum effectiveness.

- Leaking or nonfunctional septic beds create nutrient and bacterial pollution.
- People living in the country or at cottages get their drinking water from wells or sometimes lakes. If these sources are contaminated from septic beds or farms, people can get very sick.

## Fishing and Hunting

- Lead poisoning from spent shotgun pellets is a major cause of mortality in ducks, geese, doves, and other species. Birds eat small pebbles and grit, which stays in their gizzard to aid in digestion. Unfortunately, lead shot is frequently ingested also, affecting the birds' nervous systems. This can eventually cause death. Use steel shot instead.
- Lead sinkers used for fishing are another major source

of lead poisoning in larger birds, such as loons and swans, which accidentally swallow them. Loon populations already in decline in many parts of North America because of pesticides, are now threatened by lead poisoning. Use environmentally friendly sinkers instead.
- Many states and provinces are making steel shot mandatory, especially for waterfowl hunting.

## ENVIRONMENTALLY FRIENDLY CLEANERS

Everyday, we can all take simple steps to help reduce our environmental impact. We can minimize our use of hazardous household and garden products, and be sure to dispose of them carefully. We can wash with phosphate-free soaps and detergents. We can purchase products that have little or no packaging and reuse our own shopping bags. By composting kitchen vegetable wastes, we reduce our garbage by at least 25 percent. If we also recycle paper, plastic, metal, and glass, we can eliminate all but a very small amount of the remaining household waste. In the garden, we can compost leaves; we can also apply natural fertilizers, avoiding times of heavy rainfall. And using less water — in the home, garden, and workplace — helps to reduce water pollution and conserves a precious resource.

The 3 Rs of reduce, reuse, and recycle are all simple and environmentally friendly suggestions which usually save money, too. Another positive thing we can all do is to use "eco-alternatives" to harsh, commercial household cleaners and garden pesticides. They are just as effective and usually cost a lot less. Did you know that with baking soda and white vinegar, you can clean just about anything? Or that in the garden, ants and aphids steer clear of mint plants?

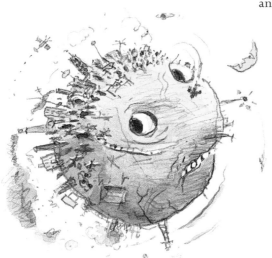

The suggestions which follow are just a start. Contact local environmental groups, government offices, libraries, and garden stores for more information on alternatives to toxic chemicals. There is a wealth of free information available in the form of booklets, fact sheets, and the like.

# Alternative Cleaners

### All-purpose cleaners

Mix water and baking soda to make a scouring agent for sinks, toilets, ovens, and pots. If pots have baked-on food, add boiling water and let soak for a while.

Combine a little detergent with baking soda to clean greasy surfaces.

### Air fresheners

Place a little vinegar or baking soda in a dish to absorb odors, or keep sachets with spice or potpourri. Keep a box of baking soda in the refrigerator and freezer.

### Bleach

Instead of chlorine bleach for cleaning, use baking soda or borax. Instead of bleach for laundry, use washing soda or borax.

### Clogged drains

To avoid clogs, keep solid wastes, fats, and hair from going down the drain. Use drain screens in kitchen sinks and showers, and occasionally pour a little salt and boiling water down the drain. If you get a blocked drain, use a plunger or plumber's "snake" to loosen the clog. Then, pour about 2–4 tablespoons (30–60 mL) baking soda and ½ cup (125 mL) vinegar in the drain, cover and wait about fifteen minutes. Finally, pour boiling water down the drain. If necessary, keep plunging!

### Detergents and soaps

Always use phosphate-free products.

### Floors

Sponge-mop floors with a solution of ½ cup (125 mL) vinegar and 1 gallon (4 L) water.

### Furniture polish

On woods finished with varnish, lacquer, or shellac, dust with a damp cloth and wipe dry. Or prepare your own furniture polish in a spray bottle with a mixture of 2 tablespoons (30 mL) olive or vegetable oil, 1 tablespoon (15 mL) vinegar or lemon juice, and 1 quart (1 L) water. Afterwards, you can sprinkle on cornstarch and rub to get a high polish.

For unfinished woods, prepare your own polish in a spray bottle with a mixture of one part lemon juice and two parts olive, vegetable, or mineral oil. You can also combine 1 tablespoon (15 mL) lemon oil with 2 cups (500 mL) mineral oil. Simply spray mixture on and polish dry.

### Glassware, mirrors, and surfaces

Combine 2 tablespoons (30 mL) white vinegar with 1 quart (1 L) water in a spray bottle. Spray on and wipe off with newspaper.

### Grease spots

Wipe with baking soda or mild detergent and water. Add a little salt to greasy spots, then wash off or scrape away residue when dry (in ovens, for example).

### Indoor ants

If you are plagued by ants inside the house, determine where they are getting in and block their path with a line of chalk, paprika, chili, or cayenne pepper. Keep your kitchen really clean and store foods in airtight containers; wash surfaces with a strong vinegar solution.

### Sanitize surfaces

Wash surface, then wipe with a strong vinegar solution and a clean cloth, or use borax.

### Spray starch

Combine and mix thoroughly 1 tablespoon (15 mL) of cornstarch with 2 cups (500 mL) water. Spray on and iron.

### Stain removers

Scrub stain with a wet paste of water and baking soda (or cornstarch).

### Toilet bowl cleaner

Sprinkle baking soda in bowl and scrub with toilet brush.

Remember: NEVER mix chlorine bleach with other chemicals, including toilet bowl cleaners, ammonia, rust removers, or acids (such as vinegar or lemon juice) because poisonous gases are formed.

# 7
# THE IMPACT
# OF EXOTIC SPECIES:
## Hidden Invaders

While people may argue about the long-term effects of chemical pollutants on the health of natural systems, the actual root sources for these pollutants are usually quite apparent. This holds true whether one pictures a factory belching smoke into the atmosphere, cattle defecating into a clear headwater stream, or the cumulative effects of thousands of city dwellers cleaning their toilet bowls with toxic chemicals.

There is, however, another form of pollution that is far more subtle. This is the issue of exotic, or alien, species. These are plants, animals, and other creatures that become established in areas where they were not formerly found. Some of these alien species were introduced accidentally by people, whereas others have been intentionally released into different habitats.

Exotic species are found in both aquatic and terrestrial areas in probably every single region on the planet. Many nonnative species are unable to survive in new habitats, and others may persist in low numbers. But not infrequently, the absence of natural predators and competitors, coupled with high reproductive potential, favors the newcomers and their populations increase dramatically. Even if their numbers stabilize, these invasive exotic species can have extremely harmful, far-reaching effects for native ecological communities. Some exotics become so abundant that they reduce the numbers and variety of native species, and therefore have a detrimental effect on biodiversity.

Aquatic ecosystems provide some dramatic examples of the impact of exotic species. Across North America, people have been stocking lakes and rivers with nonnative species of fish. In many cases, fish communities are almost entirely artificial. While this may provide recreational opportunities for some, many native fish and wildlife species have declined precipitously as a result. This, in turn, has had a negative impact on other fisheries.

A further aquatic example is that of the zebra mussel. Since its discovery in Lake St. Clair in 1988, this thumbnail-sized invertebrate has expanded rapidly through the Great Lakes–St. Lawrence watershed, and has now infiltrated the Mississippi system and others. Zebra mussels can attach themselves to virtually any hard (and even some soft) surface and have threatened or eliminated a great many native species. They feed heavily on the microscopic algae at the base of the food chain, and therefore compete with organisms in the established natural food webs.

Exotic species can have similarly devastating effects in wetlands and on land. Again, the general pattern is the same. Once firmly established, aliens can become widespread and cause the elimination of native species through competition or predation. Some nonnative species, like purple loosestrife and Norway maple, become so abundant that they form vast regions of just their single species. These monocultures reduce the biodiversity of not just plants, but of all the countless native invertebrate and vertebrate species that were associated with the habitat before the exotic species arrived.

Throughout the world, the introduction of exotic species has had a dramatic impact on native ecological communities. Some native wildlife and plants have suffered, populations have declined, and species have become endangered or extinct as a result. While species do expand and change their ranges on their own, humans have, to say the least, accelerated this process tremendously – both by accident and design. Most of these introductions would likely not occur in tens or hundreds of thousands of years, if at all.

## Zebra Mussels

- Zebra mussels arrived in the Great Lakes watershed from eastern Europe, carried in the ballast water of ships during the late 1980s.
- These coin-sized invertebrates breed prolifically and can cover underwater surfaces with a blanket of sharp, striped shells. They can completely block water-intake pipes and make swimming very unpleasant.
- Zebra mussels have a dramatic impact on aquatic ecosystems. They filter-feed plankton from water, removing nutrients that would otherwise be used by zooplankton and fish. They also grow on top of native clams and invertebrates, which can ultimately kill the native species.

## DECEPTIVE BEAUTY: PURPLE LOOSESTRIFE

Purple loosestrife is a deceptively attractive exotic plant which grows best in shallow wetlands. Its tall stalks are becoming an increasingly common sight in marshes, wet meadows, pond and river edges, and roadside ditches.

As with many exotics, the lack of natural predators helps the species to spread rapidly. Purple loosestrife can produce millions of seeds a year; they are most frequently spread by air, water currents, and the transport or movement of soils.

Once established, loosestrife quickly dominates the wetland and chokes out native species. The generally more diverse native plant community is lost, and wildlife reliant on these plants can also be lost or at best displaced. Over time, the wetland may actually dry out as a result of the proliferation of loosestrife.

## SEA LAMPREYS IN THE GREAT LAKES

Sea lampreys have a rasping, suctionlike mouth that clamps onto the flesh of large fish. Once attached, they literally suck the lifeblood from their unfortunate host.

Lampreys entered Lake Erie and the upper Great Lakes through the Welland and other ship canals. The extensive St. Lawrence Seaway connects the Atlantic Ocean to inland ports as far away as Lake Superior, and makes travel between the lakes possible for aquatic species.

The establishment of this invasive exotic was a major factor in the near complete collapse of lake trout populations, and has harmed commercial fisheries for whitefish. Control measures are in place to keep the lamprey's population in check.

**LAKE TROUT WITH PARASITIC SEA LAMPREYS**

# NORWAY MAPLES AND THE NATIVE FOREST

Norway maple was introduced to North America in the late 1800s for landscaping. It has since become widely "naturalized" and can replace native forests.

Like many invasive exotic plants, it produces large quantities of seeds. Norway maples create very heavy shade conditions, which are unfavorable for many native species of trees, shrubs, and wildflowers.

In some areas, Norway maples form a near monoculture forest, with little or no understory or groundcover plants. The resulting loss of biodiversity affects all

**Native Forests**

Native forests usually contain a rich diversity of trees and other plant species. The largest trees form the forest canopy, with lower vegetation forming the understory and groundcover. This diversity of plant material provides habitat for a wide variety of wildlife species.

## EXOTIC SPECIES

- If you must plant non-native species, check first to see if they have the potential to become a problem if they spread from your garden. Purple loosestrife, for example, is still sold by some nurseries despite its notorious ecological record.

- Avoid accidentally spreading aquatic exotic species. If you transport a boat (including canoes) from one lake to another, check to make sure you're not accidentally carrying any "hitchhikers." Scrape zebra mussels off your boat hull, and empty and clean out all bilge water.

The larval and adult forms of many invasive species are too small to be seen by the naked eye.

- Do not let pets (cats, dogs, turtles, fish, etc.) go in the wild, where they can become predators or competing species for native wildlife. Find a suitable home for them, and think twice before making a pet purchase.

- Discourage government agencies and outdoors groups from stocking non-native species of fish and game in rivers and lakes.

- Heavy shade from the Norway maples means that few, if any, plants can grow underneath. This type of forest contains little biodiversity

# 8

# HABITAT
# LOSS AND
# DEGRADATION

The loss and destruction of natural habitats is one of the greatest environmental insults that humans have inflicted upon the earth. Whether through urbanization, agriculture, or the damming of rivers, the environmental effects of habitat degradation are far-reaching and ever increasing. Living things need places that are free of pollution, and that contain the sort of habitat for which they are adapted. When forests are cut or native prairies are plowed under, or when roads are constructed, native flora and fauna are bound to suffer. We must continually take measures to reduce the damage that has been done to the natural environment, and restore the natural habitats that have been disrupted.

The timeline on page 105 lists the typical habitat changes that have occurred over the last three hundred years or so in forest and river habitats. The native forest is cleared for farming and towns are built near the rivers, often to the detriment of wetlands. As population increases, more forests are cleared, towns swell into large cities, and suburban sprawl gobbles up agricultural land. This typical pattern is occurring everywhere – and in all sorts of habitats, from native prairie and eastern forests to deserts. Every once in a while, though, the trend is reversed when marginal farmland is abandoned and natural succession lets the wild lands gain a new foothold.

## TYPICAL HABITAT CHANGES

**1.**

Mature native forests.
Marshes at river mouths
and along banks.
Small, scattered human
population.

**2.**

Forests cleared for farming.
Towns and cities developing,
often near rivers and lakes.
Marshes filled in and aquatic
habitats lost.
Increasing human population.

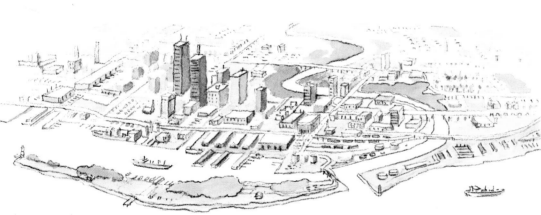

**3.**

Cities, suburbs, and industry
expanding rapidly.
Farmland near cities being
lost to urban sprawl.
Only small, scattered patches
of forest remain near cities.
Natural river and shoreline
ecosystems within cities lost
and polluted.
Human population growing
rapidly, increasingly urban.

# Meddling With the Flow of Water: Dams and Other Intrusions

Day after day, water continues its tireless travels through watersheds and around the globe. This natural hydrological cycling continues, but through our collective actions, humans have interfered with the process.

The most dramatic and obvious physical intrusions on the water cycle are dams. Historically, dams were used as a source of energy for grinding wheat and cutting logs. Over time, dams have become increasingly numerous and monstrously large. Dams do provide clean hydroelectrical power without releasing air pollution or radiation, and they make it easier to redistribute water for crops. Dams can also be used for flood control, but this requires careful monitoring; not infrequently, human error combines with natural storm conditions, and devastating flooding occurs. Recent and fatal examples can be found from across North America, including the Mississippi, Saguenay, and Red River floods.

Dams, however, do not come without significant environmental costs. Dams completely change the aquatic ecosystem: what was once a river ecosystem becomes a large lake ecosystem. This affects all of the species adapted to flowing water and the oxygen-rich conditions before the dam was built, and also has an impact on species along shorelines and on land. Once the dam is built and floodwaters have risen, plant materials decompose and rot. This reduces oxygen concentrations, particularly in the deepest parts of the reservoir. This problem can be acute in summer, when surface waters, which absorb oxygen from the air, do not mix with cold, deeper waters. Methyl mercury poisoning in fish, wildlife, and people can be a major problem in new dam impoundments. In addition, dams block traditional migration routes for fish, and trap nutrients upstream, thereby reducing their availability downstream.

Humans have caused other major environmental changes that relate to the water cycle. For example, the clearing of large expanses of forest has a big impact on entire watersheds. The removal of vegetation from upland and riverside areas increases the risk of flooding, since plants and wetlands absorb large amounts of water. By retaining large amounts of vegetation, there is also reduced runoff, less

**Climate**

● When large amounts of vegetation are removed, soil dries
out and surface runoff quickly
disappears. The lack of trees
also decreases local humidity
levels because trees release a
lot of moisture through transpiration from their leaves.
● The entire region becomes
increasingly dry or arid, and local
rainfall downwind is reduced.
● In turn, local streams and
rivers carry less water, and the
level of groundwater falls.

erosion, and more water infiltration into the soil. This is good for soil organisms
and tree roots, and helps to replenish the water table.

The loss of large amounts of vegetation within a watershed can also affect
climate. With little groundcover, soil dries out more quickly and water runoff
becomes more rapid. The local microclimate becomes less humid, which affects
the diversity of native species. On a large scale, regional weather can also be
affected by the loss of large amounts of vegetation, and many areas experience
less rainfall as a result.

Underground aquifers are essential water resources for people and nature
alike. When too much water is withdrawn for cities, agriculture, or industry,
these precious reserves can be depleted and major problems occur. Local supplies fall short and creeks, springs, and wells for miles around can go dry. When
this happens, people must truck water in or build pipelines, but the local plants
and wildlife are left dry.

Many rivers are used as a source of water for irrigating crops, nourishing
cities, and turning natural desert habitats into grassy lawns. The Colorado River,
for example, is pumped nearly dry by the time it reaches its estuary on the Pacific
Coast. The image of a river not being able to complete its voyage from headwaters
to natural outflow is pathetic. The lack of fresh water and nutrients also affects
the health of the estuary environment downstream. Water withdrawals and
aquifer depletion of this scale are not sustainable. Perhaps we should rethink
where we grow crops that require heavy irrigation or where we locate major
new cities. After all, there is a reason why drought-resistant plants grow in dry
climates, and those requiring lots of water grow in rainforests.

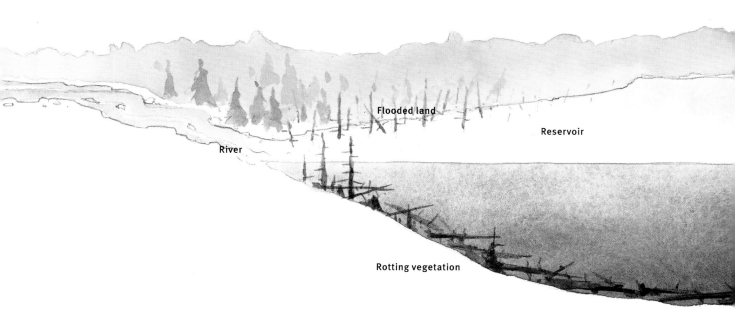

Flooded land

Reservoir

River

Rotting vegetation

## DAMS

- Dams block the migration of fish, such as salmon, trout, and eels.
- Species adapted to flowing water and the high oxygen levels of rivers are displaced.
- The decomposition of wood and vegetation at the bottom of deep reservoirs consumes oxygen. In summer, oxygen levels can become dangerously low.
- Methyl mercury poisoning of wildlife and fish is a major concern in newly constructed dams.
- Dams hold back water until it is needed to generate electricity, and so the peak in water (and also nutrient) flow in dammed rivers often comes in winter. Aquatic ecosystems are adapted to receiving nutrients and peak water runoff in the spring; thus the natural cycle is disrupted.

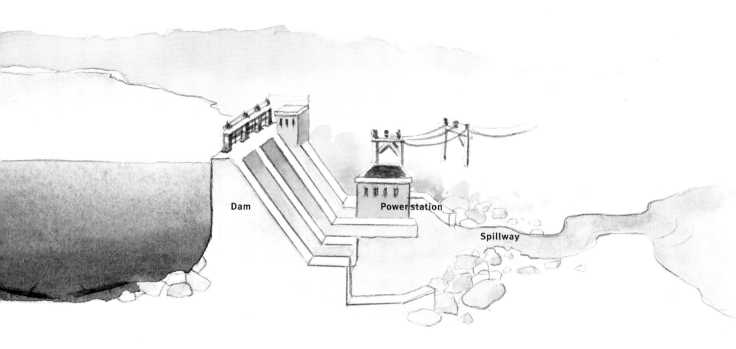

Dam

Power station

Spillway

- When hydroelectric projects are built, the size of the areas flooded is some-times enormous — larger even than some countries. This has a dramatic impact on habitats.
- There are hundreds of thousands of dams in the watersheds of North America. They range in size from very small to gigantic and are found near our most densely populated cities and in the most remote sections of the continent.
- Power utilities continue to plan enormous hydrodevelopment projects for remote rivers. More efforts should go into energy conservation measures before new hydroelectric developments are built.
- Keep a river wild and free, use less electricity!

## FORESTRY ISSUES

When the first waves of European settlers arrived in North America a few centuries ago, they were faced with a seemingly endless expanse of trees. With the exception of prairie and desert regions, and the very far north, almost all of the new continent was covered by forest of one kind or another. Since many of these new citizens wanted to farm the land, the landscape seemed most inhospitable. To a large extent, forests were adversaries to be conquered: clear the land, build a homestead, and start farming the soil.

For some settlers, though, the forest itself was a resource to be exploited. High-quality wood products from virgin stands of timber were highly sought after for growing North American markets and for abroad. During the nineteenth century, for example, white pine was abundant and in demand in Quebec, Ontario, and New England. The forests were abuzz with loggers, lumber camps, and the sounds of big saws and axes; and spring melt signaled the start of the exciting but dangerous season of log drives which carried squared timbers to waiting ships.

During the twentieth century, human settlement has spread into every corner of the continent and our numbers have soared. This exponential growth in population has been matched with an increasingly mechanized forest industry. As a result, forestry is now big business and carries considerable economic and political clout.

# Logging Practices

"A forest is more than a bunch of trees."

Forestry practices vary across North America, reflecting the different types of trees being harvested, the different types of terrain, and also the ecological sensitivity of the company doing the cutting. The logging industry is at the center of many environmental controversies. When complex natural forests are cut, the biodiversity of both plants and local wildlife populations is reduced. When companies do reforest an area, it is usually with very few tree species (frequently only one or two). While things may appear green again, the natural ecosystem has been dramatically altered and the natural biodiversity is reduced. A tree farm is not the same as the native forest it replaced.

In a great many cases, forests are not replanted, and natural succession must begin anew. Unless there is serious soil erosion, a new forest will gradually develop and some wildlife species will benefit in the short term. Unfortunately, clearcuts may go on for miles and miles. This means that many species of wildlife and plants are completely excluded from areas where they formerly lived. Species such as some of the big carnivorous mammals which require large expanses of forest habitat are particularly at risk from this sort of habitat disturbance.

## Mature Native Forest

● Mature native forests contain a wide variety of tree, shrub, and herbaceous plant species. "Virgin forests" are original, stable, old-growth forests that have never been cut. They contain a very rich biodiversity of animal and plant communities.

● This forest has many layers of vegetation, from small groundcover plants to understory shrubs and small trees up to the tallest forest giants.

● The diverse and extensive forest cover stabilizes soil and reduces erosion. Fallen leaves and branches add nutrients to the soil. Extensive forests help hold moisture in the soil.

● The great diversity and vertical layering of plants in the mature climax forest provides many habitats, which explains the high biodiversity of animal life.

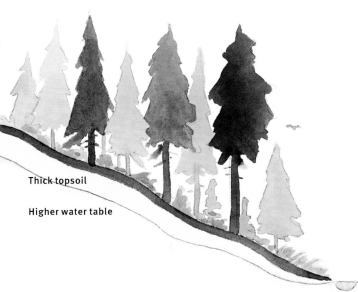

Thick topsoil

Higher water table

Stream

## Selective Logging

- In selective logging, a forester marks individual trees and then the logging crews cut them. In this case, much of the original forest remains and natural regeneration occurs. There is only limited risk of soil erosion, stream damage, and total habitat loss.
- The heavy equipment used, however, can damage trees and scrape the soil. Selective logging also removes the healthiest individuals and the most desirable species, which affects the overall species composition in the forest. And some forest companies are more "selective" than others. In general, though, this method is preferable to clear-cutting.
- Logging companies point out that selective cutting is not feasible or safe in some circumstances, depending on the steepness of slopes and the type of forest. For example, selective cutting is not usually practiced in the pulp and paper industry in the boreal forest, since these trees are small and of low market value.

Seed source from standing trees

Trees hold soil in place

High water table

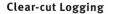

Slash piles provide nutrients

## Clear-cut Logging

- Clear-cut logging removes all of the trees from a given area.
- This method is used extensively in the northern boreal forest, where spruce and fir trees are used for pulp and paper or construction. Clear-cut logging is also practiced widely in mountainous and remote areas where operating costs are high.
- When all the trees are removed, the nutrients in the wood are lost from the ecosystem and soils become nutrient poor. In contrast, many nutrients remain after a forest fire or insect infestation.
- After clear-cutting, increased soil erosion can occur, especially if the cut is on a hill or mountainside. In some cases, both vegetation and topsoil are lost, leaving little more than bare gravel and sand, making forest regeneration difficult or unlikely.
- Without forest cover, soils become drier because of the increased evaporation and lowering of the water table. As a result, small streams in the area sometimes dry up.
- Reforestation is now widely practiced after clear-cutting. This reduces the risk of soil erosion and helps reestablish trees. In most cases, however, only one or two species are replanted, and the diverse forest community is lost.

Evaporation from soil

Lower water table

Topsoil lost

Rapid runoff

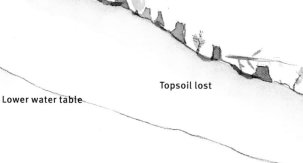

Dry, silted stream bed

## Additional Forestry Problems

Numerous other problems are related to forestry practices. These include the intensive use of pesticides to control insects and disease, large-scale erosion from logging roads and clear-cutting on steep slopes, the warming of forest streams because of the loss of shade cover, and pollution discharges to air and water by paper mills.

- Logging roads allow easy access in to remote regions, leaving increasingly few places where wildlife remain undisturbed.

- If logging occurs right up to the edges of rivers and streams, silt and soil end up in the water, which can suffocate fish and kill aquatic life. The same problem can occur where logging roads cross streams.

- Fallen logs and logging debris that gets washed into rivers can block fish migration routes.
- If there is no shade from trees, streams get too warm for some creatures.

- Logging companies should leave wide buffer edges along waterways to reduce these problems.

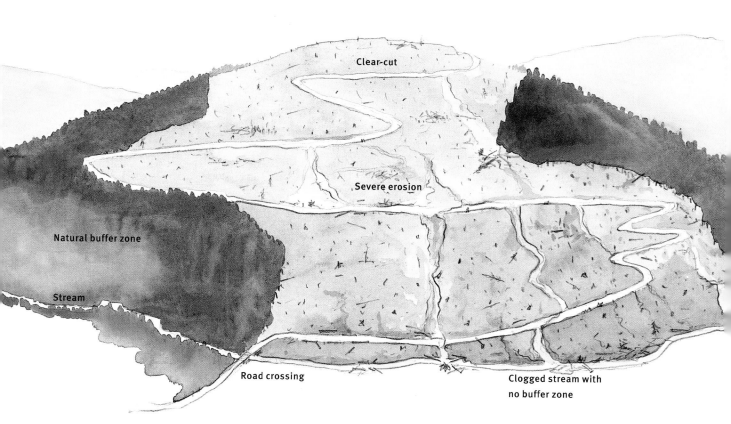

Clear-cut

Severe erosion

Natural buffer zone

Stream

Road crossing

Clogged stream with no buffer zone

## Alternative Logging Practices

- In strip-cutting, long and relatively narrow bands of trees are clear-cut and the adjacent mature forest strip is left standing as a source of seeds for forest regeneration. Forested cross-connections allow wildlife to move between the wooded sections. With this form of logging, a greater diversity of life can remain in the same area. The standing strip is usually cut later.
- "Shelterwood logging" removes some of the trees, but leaves the most highly desirable specimens standing for further growth and as a source of seeds.

## How Can I Help?

### SAVING THE FORESTS

**For the individual**

- Reuse scrap paper for notes and draft documents.
- Recycle all used paper and cardboard products.
- Use paper and wood products conservatively.
- Use your consumer power. Lobby the forest industry and governments for improved environmental responsibility. Big companies *are* responding to public pressure about clear-cutting, especially in old-growth forests.

**For the forest industry**

- Cut selectively.
- Stop clear-cutting on steep slopes.
- Reduce the size of or phase out clear-cuts.
- Improve road crossings over streams to reduce erosion.
- Leave wide natural buffers around streams and lakes.
- Use alternative crops, such as hemp, for fiber.

# AGRICULTURAL AND RURAL ISSUES

*Endless fields of wheat and hay*
*bow rhythmically to breezes*
*while cattle graze placidly*
*beside a country stream.*

Pastoral images of rural landscapes are undeniably appealing, but they are as misleading as they are innocent. Many of us assume that what we see in rural areas is representative of a natural landscape, and that it is therefore an excellent habitat for wildlife. Unfortunately, any region that is being used intensively for crops or livestock is very different from the original ecosystem located there. Across the continent and around the world, enormous tracts of land have been converted from native forests, native grasslands, and even deserts, into regions of food production. The lush green of farmland is misleading, at least in terms of natural habitats.

When the settlers cleared sections of forest in eastern North America, they replaced the native plant and animal communities with crops and livestock. On a small scale, the effects were minimal. Initially, displaced wildlife could move on, and the small number of settlers meant that there was still plenty of native forest left. But more people arrived and more were born, so the demand for agricultural land kept increasing.

**Native Ecosystems versus Monoculture Crops**

• Forests and grasslands have many different species of plants in them, organized into many different layers. Some plants grow tall, some to moderate heights, and some are low to the ground. This vertical stratification, combined with species diversity, provides innumerable microhabitats for all sorts of organisms.

• Compare this wonderful diversity to what is found in a single crop.

As a result, huge areas which were formerly forest are now fields of crops or pasture. In some regions, the land was cleared two hundred years ago, leading many people to believe that these regions were always open country. Many wildlife species that were adapted to the original forest habitat are now much less abundant, found only in fragmented patches of their native habitat.

This same story was repeated across the continent, as settlers moved west. The increasing demand for crop and pasture land meant that any area with suitable climate and soils was tried. The prairie grasslands were particularly attractive to homesteaders, since they did not have to clear the land. This change from native grassland to crop or pasture land is subtle. To many people, there is no apparent difference between a field of canola and the native grassland. The original prairie contained a great diversity of plant and animal species. When these areas were converted to single crops, or to a lesser extent to pasture land, the native species and communities were affected.

The effects of removing established plant communities are far-reaching. Clearly, the diversity of plants is immediately reduced. But, in addition, there is the resulting decrease in the diversity of animals and other creatures which relied upon the original native plants. The change from native forest to agricultural field provides the obvious example, but change from native prairie is just as significant. In fact, the native tall-grass prairie is among the most threatened of habitats in North America.

● From a distance, fields of crops, natural grasslands, and forests all appear green. They all help produce oxygen and reduce greenhouse gases through photosynthesis. In addition, the roots of plants decrease erosion, and shade from leaves reduces water evaporation from the soil. All of these things are good.

● There is a big difference, however, in how useful this vegetation is to native wildlife. The large expanses of monoculture crops so common today provide very little habitat diversity. Consequently, the biodiversity of plants, animals, and even microorganisms is greatly reduced with intensive agriculture.

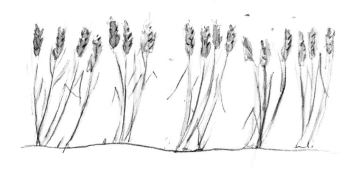

## Prairie Sloughs

- The "pothole" sloughs of the prairies are among the most important breeding grounds for ducks in North America. Like other wetlands, much of this habitat has been drained, filled in, and farmed to get a few more acres of crop production. Sometimes, the land remains too wet for good crop growth.
- These wetlands are a very important habitat for waterfowl and other animals and should be left in a natural state.
- Wetlands on the farm store water, reduce the risk of flooding, and provide fish spawning habitat. Cattails and other marsh plants also soak up large amounts of nutrients and some toxic pollutants, thereby helping to purify water.

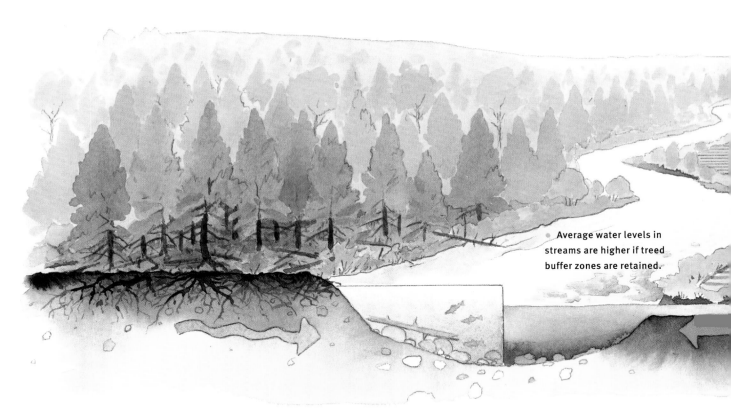

Average water levels in streams are higher if treed buffer zones are retained.

## Farming the Bottom Lands

The low, flat lands next to rivers are called bottom lands or flood plains. As the name suggests, these areas are subject to periodic flooding, usually during heavy spring rainstorms and as the result of snowmelt. Flooding brings with it large quantities of nutrients, making these areas extremely fertile. Because of the soil richness and the easy access to water, these bottom lands are desirable locations for farming, despite the hazards of flooding.

Diverse forest communities are native to many river valleys, and birds, mammals, and insects migrate along these corridors. Some types of fish have even adapted to spawn in temporarily flooded woodlands. In many areas, farms and cities monopolize this ecologically rich ground, consuming important habitat and decreasing the effectiveness of the natural corridor. As well, a lack of trees and shrubs overhanging small rivers and streams is harmful for aquatic life. Without shade, water temperature increases, oxygen levels decrease, and there are fewer places for fish and invertebrates to hide.

## How Can I Help?

### HABITAT ON THE WORKING FARM

• Retain or restore a wide buffer strip of natural vegetation along the edges of streams, rivers, and lakes to provide habitat for wildlife. This creates a natural corridor and keeps aquatic creatures healthy. Natural buffers reduce soil erosion and the risk of water pollution as well.

• Leave parts of the farm as natural habitat. Woodlots provide places for recreation, habitat for wildlife, and a free source of firewood. In many of the eastern parts of the continent, they are also a good source of maple syrup.

• Leave wetlands intact; avoid filling them in for farming.

• Let marginal farmland become wild again. These less productive lands can be left to regenerate through natural succession or replanted with native species of trees and vegetation.

## URBAN AND SUBURBAN ISSUES

Not so long ago, the majority of us lived in rural areas, but now this trend is reversed and most people live in towns and cities, close to jobs and services. The images and realities of cities are very familiar: endless expanses of houses, apartment buildings, shopping malls, factories, driveways, parking lots, roads, and expressways. These urban and suburban land uses are spreading rapidly as the population of North America continues to grow. While most urban areas have some designated green spaces, much of this land is mowed grass with just a few "lollipop" trees and precious little of the original natural habitats.

But nature can find a home in cities. We are all familiar with the urban abundance of pigeons, starlings, and racoons; given a chance, our cities can be home to many more species than that. Rare birds show up in naturalized yards and gardens in the downtown cores of our largest cities. Turtles sun themselves on logs in urban rivers. Foxes, deer, and coyotes appear in parklands and in ravines, where remnants of native forests sometimes remain uncut on the steep hillsides. Urban spaces are not suitable for all wildlife, but if there is good habitat present, a surprising diversity of wildlife species will be found – despite the close proximity of a large human population.

### Urban Tokenism

- Small urban parks with mowed grass and lollipop-shaped trees provide areas for human recreation, but are not really meaningful natural habitats.
- Sterile golf courses and urban and suburban gardens that are dominated by lawns are similarly unsuitable for most native wildlife.

## Urban Rivers

Rivers have always been important corridors for the movement of wildlife. Fish migrate up and downstream to spawn. Migrating birds follow these natural highways on their way to and from wintering grounds. And innumerable other creatures, both large and small, travel in or along rivers as part of their daily, monthly, or annual travels.

These natural river corridors also served as important routes for the travels of people: from Native Americans to Coureurs de Bois to explorers and modern-day canoeists. At one time, risk of natural flooding kept developers from building homes and businesses on the low ground near rivers. Farms and water-reliant industries were often the only human structures in the valleys.

With engineered techniques now in place, floods are less frequent. Consequently, modern urban river valleys are full of expressways, railway lines, oil and

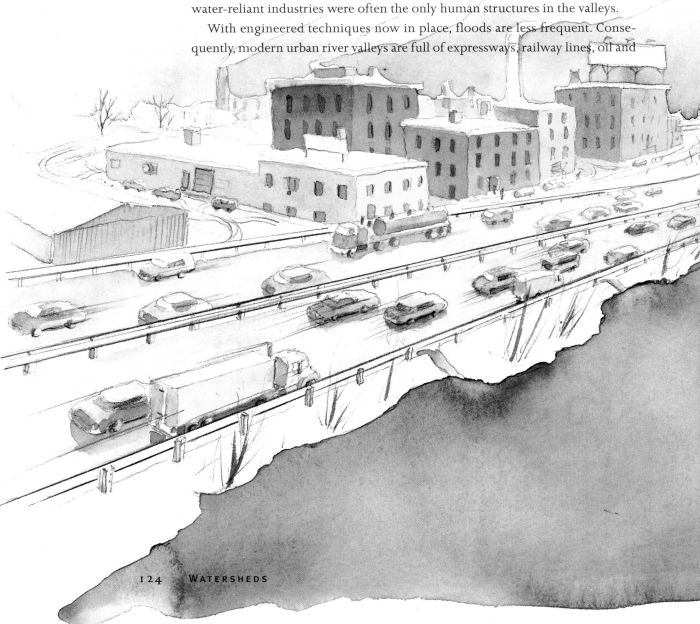

gas pipelines, hydroelectric corridors, and innumerable other nonnatural uses. In urban centers, the ecological value of rivers and river valleys has often been totally replaced by utilitarian, human-oriented functions. The greatest environmental damage and the highest density of people is usually downstream, near the mouth of the river.

In far too many places, these degraded areas provide no natural habitat at all. Lush flood-plain forests are long gone — first for farms and then for urban transportation routes. River-mouth wetlands have been filled for industry or tidy waterfront developments. Where space is at a premium, the gentle curving path of rivers has even been changed, leaving only straitjacketed prisoners to tell their tale.

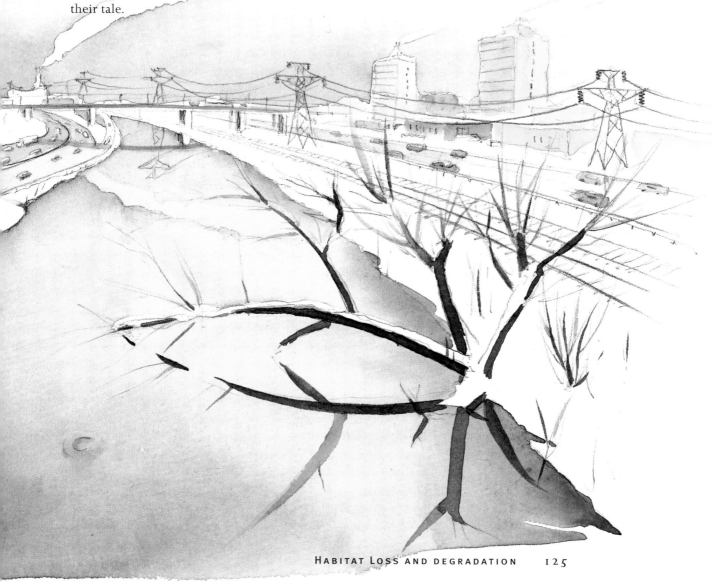

# Problems with Fragmented Habitats: Landscape Ecology

When habitat is lost or fragmented as a result of forestry, agriculture, or urban development, the ecological effects are complex. Generally, when the size of a natural area decreases, the total number of species also decreases, from plants and animals, to fungi, to bacteria and protozoans. Conversely, larger natural habitats have more species. This pattern of "island biogeography" holds true for patches of remnant forest or native prairie amid a sea of crops, as well as for real islands. The actual biodiversity of an area, therefore, is related to the size of the natural habitat.

When a natural landscape is fragmented, or broken up into smaller and smaller segments, the *quantity* of natural habitat is obviously affected. But, in addition, the *quality* of that habitat is also significantly compromised. Even a roadway that cuts through the middle of an ecologically sensitive woodland can have surprisingly far-reaching effects. This occurs because many species of wildlife do not like living near the edge of habitats.

A great many species are adapted to living in the interior of relatively uniform types of ecosystems, which is not surprising since large expanses of specific habitat types once covered the continent. Many small creatures will not cross over even a narrow roadway in the middle of an isolated northern forest. There are, however, some species that have benefited from the fragmentation of habitats. These species often prefer to live along edges.

Roads cutting through the middle of a natural area greatly reduce the amount of interior habitat and block the movement of wildlife.

## THE EDGE EFFECT

The area near the outside of the native habitat type is called the "edge," whereas the "interior" habitat is found inside. Many native species require large tracts of the interior habitat and completely avoid edges (for example, many migrant songbirds and woodland caribou).

The amount of edge habitat greatly increases as the habitat becomes more fragmented. Many parks established to protect wildlife are actually too small for interior species, and some small, remnant patches of habitat may contain all edge and no interior.

## WILDLIFE CORRIDORS

Wildlife corridors are strips of habitat that connect fragmented natural areas. They make it possible for wildlife to travel more safely between islands of habitat. Wooded river edges are a good example of a wildlife corridor. Corridors also increase the ecological value of isolated and fragmented habitat, and the wider areas (or nodes) along corridors provide more shelter, food, and protection.

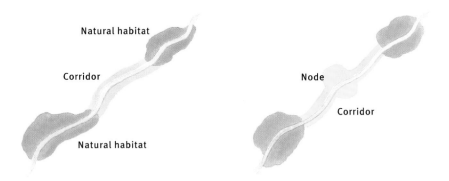

## HABITAT FRAGMENTATION

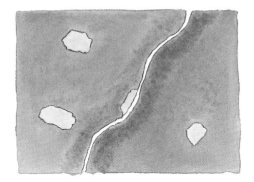

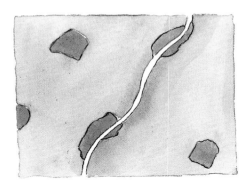

### Historic

- Original natural habitat was large and relatively uniform.
- It included some variation and water sources.
- Large blocks of habitat contained high species biodiversity.

### Present

- Original habitat is now greatly reduced and fragmented.
- Small, isolated patches of natural habitat contain fewer native species of wildlife.
- This is typical of the landscape now.

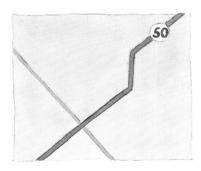

### Misleading Road Maps

- Despite the large areas of green on many road maps, ther are very few parts of North America that do not have an expansive network of roads.

- In truth, there are many more roads than most people realize. Logging and mining access roads pierce through even very remote northern and mountainous areas.

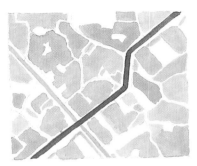

# 9
# ECOLOGICAL RESTORATION:
## Righting Environmental Wrongs

Throughout human history, we have had a profound effect on the natural ecological communities where we live. When we change a habitat to suit our own needs, we invariably affect all of the other creatures living there. In so doing, we have an impact on the entire watershed of which we are a part. Fortunately, many people are starting to realize that protecting natural habitats and native species is very important. Some new developments are incorporating a few natural features, from woodlands to ponds, and are attempting to reduce their environmental impact. And judging by the names of some new housing subdivisions, the proximity of even a *minor* green space can be a *major* selling feature.

In areas that are already developed or degraded, environmental groups are trying to restore natural habitats on vacant lots and public parklands, as well as on private property. This movement is becoming increasingly sophisticated in the ecological sense. For example, planting trees is a good thing to do, but the environmental benefit is much greater if the restored vegetation is typical of the native plant community originally found there. By linking isolated restoration projects together, these replanting efforts become even more significant. When these habitat restoration projects go hand in hand with pollution reduction efforts, the environment really benefits.

# Restoring Habitats Close to Home

With a bit of research and planning, you can landscape your yard or farm and turn it into something both aesthetically beautiful and also very inviting for native wildlife. Even if the area is small, such as a tiny downtown garden, a restoration project will be environmentally beneficial.

The first step in getting started is to find out what sort of natural habitat would have been found in your area originally, for example, prairie or deciduous forest. Generally, it makes sense to try to recreate that sort of community, since you know that it is suitable for your soil and climate. If certain habitat types in your region are particularly threatened, you may decide to help to re-establish them. Resist the temptation, though, to attempt to create a habitat that does not fit the conditions of your garden, such as trying to grow rainforest trees in a dry climate! It is always best to select something which you know to be locally suitable. In the long run, you will probably be happier if your new garden does not require a great deal of maintenance.

In designing your project, try to include species and features that could help native wildlife. For example, large trees provide shade and also nesting habitat for birds. Some small mammals and birds like dense, thorny shrubs for cover, and others skulk through low vegetation looking for food. If you include a small pond, or even a birdbath, you will be certain to attract even more wildlife, especially if you do not live near a stream or lake.

You may wish to create some diversity of plant types and sizes in your plan, which could help to attract a greater number of species. There are an infinite number of possible designs for your garden. No matter what you do, though, it will almost certainly be an improvement over a grass lawn. Consult books on garden design and guides to attracting wildlife. Check at your local garden center or contact a local natural history club or environmental group for more ideas.

# THE MULTIPLE BENEFITS OF NATIVE PLANTS

It seems far too simple to be meaningful, but planting native trees, shrubs, and other vegetation is one of the most effective ways to restore natural habitats and to improve water quality. Since plants are the foundation of food chains, restoring natural wildflowers, shrubs, and forest cover is a critical part of any effort to restore habitats. The benefits are many.

● Shade provides relief from the heat for people and animals, keeps the local climate cool, and reduces water loss from soil. This can in many places help to keep the water table closer to the surface.

● The shade of trees can also cool stream and river water, thus increasing oxygen levels for trout and other creatures.

● Plants consume carbon dioxide, decreasing atmospheric greenhouse gases and combating global warming, and they release oxygen into the atmosphere.

● Trees intercept rainfall and soak up water, reducing erosion and the risk of floods.

● Smaller vegetation and ground cover provides habitat for creatures and microorganisms.

● Vegetation soaks up excess nutrients, fertilizers, and even some toxins, reducing the risk of water pollution.

● Forests and wetlands act like sponges, holding excess water and slowly releasing it to streams and ground water reserves.

● Plant roots hold the soil in place and reduce erosion.

● Fallen leaves, branches, and dead tree trunks rot and replenish the soil with nutrients.

● Vegetation provides shelter for wildlife, nesting sites for birds and squirrels, and food for all sorts of creatures, large and small.

● Trees, shrubs, and ground vegetation mean not only more plant life, but also a much greater total biodiversity than would be found in a grassy lawn.

### RESTORING BACKYARDS

- Decide in advance what your goals are. If you are hoping to attract certain species of birds, find out what they like to eat and where they nest. This will increase your chance of success.

- Plant only native species of trees, shrubs, and small herbaceous plants. Include species that provide food and cover for wildlife. Check field guides to determine which species are native and which are alien.

- If your neighbors have a naturalized section in their garden, or if you are next to a park, make sure the naturalized part of your property is adjacent. This will help to increase the habitat's size and ecological benefit.

- All living things need water. By including a pond or birdbath, you will certainly attract more wildlife.

- Most creatures do not like stagnant water, so consider buying a circulating pump for the pond and bird bath. A raised bucket with a small hose can be also used to drip water slowly into a birdbath, or you can simply change the water daily.

- Set lawn mower blades high (2 to 2 ½ inches / 5 to 6 cm) and leave cuttings on the lawn

- Encourage birds year round. Install birdhouses for summer visitors and a variety of feeders for winter birds.
- Check with other family members to find out what they would like in the garden. If there is demand for it, leave some open grassy space for recreation.
- Water your newly planted specimens a lot at first to help them establish a good root system. You may need to fertilize them also, but be sure to use natural products. Fallen leaves and household compost make great fertilizer and act as a mulch which conserves moisture in the soil.
- Once your garden is established, you will not need to do much maintenance. Sit back and enjoy. Keep a pair of binoculars handy so that you can witness nature creeping home.
- Spread the word: restore parts of your farm, your business, your school, and your local park.

- Planting trees, shrubs, and wildflowers is a fun activity at home, at school, at work, or as part of a community-enhancement project.
- Trees and vegetation along riverbanks and lake edges are especially important.
- Be sure to plant trees at the right depth. The hole should be large enough to accommodate all of the roots, but not much deeper. Spread wood chips or other mulch around the tree to inhibit the growth of weed competitors.
- Water newly planted trees at least once a week during the first growing season, or if they start to wilt during hot and dry weather.

## STREAM AND RIVER RESTORATION

Restoring the ecological health of streams and rivers may seem like a very ambitious proposition, but the positive impacts are tremendous. Some of the major problems include the loss of riverside vegetation and forest cover, erosion of banks, stream channel widening, siltation, water pollution, and damming. Waterways that have suffered severe environmental damage may require extensive project planning and hard work, plus financial backing, to achieve major restoration goals. There are, however, many beneficial things which can be done locally that are much simpler and less expensive.

One of the easiest and most important things you can do to help a river is to plant native vegetation along the banks, which will immediately stabilize the soil. This, in turn, will result in less erosion of the banks, less sedimentation, a reduction in the amount of excess nutrients, and a reduced risk of flooding. In choosing plants, consider which species are best suited to the local growing conditions. Increasing the diversity and amount of plant life will automatically increase the variety of animal life, since many habitats will be created. To increase aquatic habitats, remember that most streams usually have *some* logs and natural debris, either in the water or overhanging it.

Highly degraded rivers, with few trees and eroded banks, typically become wider and shallower, and lose much of their natural flow pattern. With greater bank stability, the river channel may start to return to its original, narrower width. As restoration efforts proceed, much of the original flow pattern of fast-flowing riffles followed by deeper and calmer pools will also return. Once the vegetation is established, it will shade the water, keeping the temperature cooler and raising the concentration of oxygen.

Where erosion is severe, it may be necessary to reinforce the banks with logs or other obstacles which would help to decrease the current along the bank. Under extreme cases, reinforced gabion rock baskets (or rip-rap) may be necessary. To help deepen the river channel, "digger logs" could be installed to force water down. If you can get some major backing to do large-scale restoration, you may want to explore the possibility of removing dams that block the migration of fish. This is not always possible, though, because many dams are maintained for flood control.

Stream restoration efforts will be much more successful if you involve other people in the surrounding watershed. For example, encouraging local home-owners, farmers, and business leaders to become more environmentally responsible will help your efforts by reducing pollution and improving habitat. Within the stream itself, you can remove garbage and old tires, which are the source of many toxins.

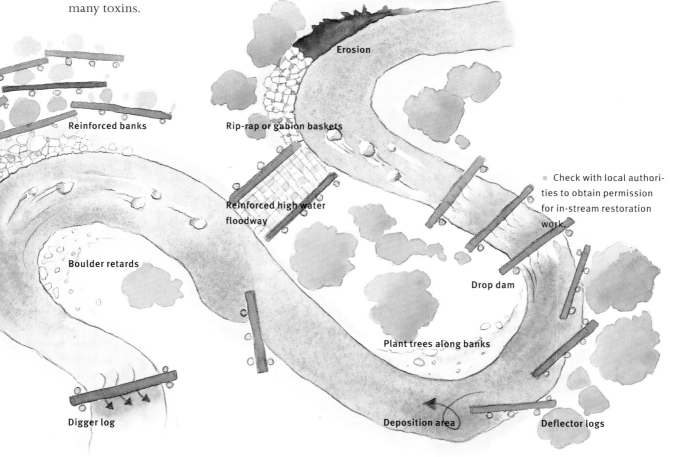

Erosion

Reinforced banks

Rip-rap or gabion baskets

Reinforced high water floodway

Check with local authorities to obtain permission for in-stream restoration work.

Boulder retards

Drop dam

Plant trees along banks

Digger log

Deposition area

Deflector logs

# ASSESSING YOUR LOCAL STREAM: HOW CLEAN IS THE WATER?

## Biological Indicators of Water Quality

You can get a good indication of how clean your local stream or river is by taking a quick look at some of the aquatic creatures found there. Many organisms can live only in water that contains lots of oxygen (which is usually the case in clean, fast-flowing streams and rivers). Some species can tolerate slightly polluted waters, whereas others will live in areas that are highly polluted. Species that reflect the health of the environment are called biological indicators, bioindicators, or bioindex species. Pollution-tolerant species can be found in cleaner waters, but clean water species will not be found in highly polluted habitats. Generally, as the amount of pollution increases, the number of species decreases.

No special equipment or expertise is required in this biological assessment: rubber boots and a collecting container will do. If you want to get fancy, bring along a dip net, forceps (tweezers), a magnifying glass, and a white enamel pan. Quietly observe the water for a few minutes and you are bound to see things moving. You should also look at the underside of stones to check if there is anything crawling around; examine vegetation and probe the bottom sediments, too. If you put a few small algae or moss-covered stones and some rotting leaves in a clear container, you may see even more things starting to move.

The list on page 139 represents a generalized summary of pollution tolerance for different groups of aquatic invertebrates, plus a few fish and amphibians. These species primarily reflect the amount of oxygen present, which is an essential aspect of the aquatic ecosystem. Indirectly, though, this can also indicate the amount of pollution from nitrogen, phosphorus, agricultural runoff (including manure), and even sewage.

This listing is not intended to indicate the amount of pollution from toxic chemicals. The presence of "clean water" fish and amphibians, however, does provide some positive clues. The eggs of frogs, toads, and salamanders are laid, hatch, and develop in water, and they are highly sensitive to pollution of all sorts, including acid rain, heavy metals, and other toxins. This means that they are very useful, albeit rather unfortunate, bioindicator species. If they are absent from suitable habitat, it is possible that there is a pollution problem, although other explanations (such as hunting and collecting) are also possible.

# Bioindicator Species for Fast-Flowing Streams and Rivers

**FOUND ONLY IN CLEAN WATER**

Mayfly larvae (Ephemeroptera)
Stonefly larvae (Plecoptera)
Dragonfly larvae (Anisoptera)
True bug larvae (Hemiptera)
Beetle larvae (Coleoptera)
Caddisfly larvae (Trichoptera)
Freshwater sponges (Porifera)
Trout and salmon (Salmonids)
Frogs, toads, salamanders
(Amphibians)

Trout

**CAN LIVE IN SOMEWHAT POLLUTED WATER**

Leeches (Hirudinea, Annelida)
Aquatic sowbugs (Isopoda)
Crayfish (Decapoda)
Some snails (Gastropoda)
Some midgefly larvae
(Chironomidae)
Carp, minnows, goldfish (Cyprinidae)
Suckers (Castomidae)
Catfish, brown bullhead (Ictaluridae)

Crayfish

**CAN LIVE IN VERY POLLUTED WATER**

Sludge and segmented worms
(for example, Tubificidae)
Midgefly larvae (Chironomidae,
including red larval
"bloodworms")

Midgefly larvae

- Remember that this list of bioindicators is a guide to the environmental health of flowing water ecosystems only, such as streams and rivers. It is not intended to be used for lakes, ponds, and wetland ecosystems, which often have very different aquatic life.
- The illustrations in Appendix III will help you to identify these common aquatic creatures.

## A Word about Safety . . .

A few precautions before you get your feet wet: make sure that stream and river conditions are safe. Avoid deep or treacherous local areas, including areas with a lot of sticky clay or other hazards. During flooding, rivers and even tiny streams can rise quickly and become very dangerous. Stay in shallow, rocky areas. These are the best areas to look for bioindicator species. If you suspect heavy contamination with harmful bacteria (from sewage or livestock waste), check with local health officials and wear rubber gloves. Wash your hands after your investigations.

# CONCLUSION

This book has explored some of the fundamentals of ecology and also a great many environmental issues. Throughout this discussion, we have emphasized the importance of water and watersheds to wildlife, to ourselves, and to the general health and well-being of the natural world. Too often, people have forgotten a very simple and important point about ecology: all living things need water. And that water is needed in abundance, and it is needed clean and fresh. Water, therefore, sustains life on this incredible planet.

North America is rich in natural resources. There are expansive forests and fertile agricultural lands, and in some areas there are wide open spaces where the human population remains relatively sparse. In many parts of the continent, fresh water is still found in great abundance. It is, in fact, so freely available in areas near large lakes and rivers that people take it far too much for granted.

Despite the importance of this most basic resource, humans have squandered and polluted water. We have used waterways as highways of exploration and for transportation. We have harnessed the currents to float log drives, to produce energy for grinding wheat, and to cut timber. We block and change the flow of rivers large and small in order to generate electricity. We withdraw water with reckless abandon from surface and underground reserves to grow crops where desert plants alone once grew. We remove or change forests and other vegetation so much that the flow of water and the replenishment of hidden aquifers are diminished. And for centuries, we have built our farms, villages, towns, and cities where fresh water could provide for us. Then, we use these same rivers and lakes to dilute and carry away our sewage and other household and industrial wastes.

It is heartening, then, that we have come a very long way in recent decades in our approach to water. People are recognizing the problems, and many positive steps are now under way to improve air, soil, and water quality. Some of these actions are massive, such as the cleanups of highly contaminated sites and the installation of more environmentally friendly technologies. A great many – perhaps a great majority – of the positive steps being taken, though, are small scale, local, and low budget. Teams of volunteers pick up garbage, plant trees, or help to re-establish long-forgotten wetlands. On an individual basis, people are taking progressive ecological steps – on farms, in factories, and in homes. We *are* restoring habitats and we *are* taking steps to reduce air and water pollution.

By considering the impact of our actions on watersheds, we understand that everything that happens in the air and on the land will ultimately affect the water. We know, too, that problems upstream accumulate and build with each tributary stream. We know that these environmental issues become magnified both downwind and downstream. But our positive actions, too, can become magnified. Ecological restoration and pollution cleanups have a beneficial impact, not just locally but throughout the whole watershed. These actions may seem small sometimes, but when millions of us take part, the benefits are great. So next time you do something that helps our watersheds, please do one more thing – get a few friends and relatives to join in, and build on something good.

We are all drawn to water with irresistible attraction, and we all seem to share a childlike fascination with streams, rivers, lakes, and shorelines. Perhaps this is because water is always on the move, and because it is so vital to life and to ourselves. We should be fascinated with meltwater streams, with surging rivers, and waterfalls, but also with bogs and swamps. This inexplicable and natural attraction sometimes tempts us to follow a stream from its humble headwater origins all the way downstream to its outflow in the sea. It should be equally natural then, too, for each of us to do all that we can to ensure the well-being of waters and watersheds, no matter where we live.

# Appendices

## I     The Five Kingdoms of Living Things

*Monera*

*Protista*

Scientists who name species and who study the relationships between different groups of organisms are called taxonomists. They classify living things into the following five kingdoms: Monera, Protista, Fungi, Plantae, and Animalia. This Five Kingdom Classification replaces an earlier and very well known system which divided all living things into either the plant or animal kingdoms. The organisms within each of these five distinct classifications share specific characteristics.

To avoid possible confusion regarding common names, all known species are given a globally accepted and standardized scientific name. This name consists of two essential parts: the genus, which is listed first, and the species, which is second. For example, the scientific name for the peregrine falcon is *Falco peregrinus*. No two species have exactly the same scientific name; closely related species, however, often have the same genus name.

### MONERA (BACTERIA)

Monera are single-celled, and are the tiniest and most ancient of living things. They first evolved over 3.5 billion years ago. (The age of the earth is about 4.6 billion years.) This kingdom includes species that are helpful (e.g., oxygen-producing cyanobacteria or the bacteria needed to produce cheeses and yoghurt) as well as harmful (e.g., disease-causing bacteria, such as pneumonia or botulism bacteria).

### PROTISTA

This kingdom includes many single-celled organisms such as amoebas and paramecium. Taxonomists now include the large, multicellular seaweeds and kelps in this group.

### FUNGI

Members of the fungi kingdom are usually multicellular. They include molds, mushrooms, and many decomposer species. The mushroom cap is the reproductive part of the fungus; most of the fungus consists of hidden and transparent hairlike threads.

### PLANTAE

Members of the plant kingdom are multi-cellular organisms that make their own food through the process of photosynthesis. This group includes mosses, grasses, ferns, wildflowers, and trees, to name just a few.

### ANIMALIA

Members of the animal kingdom are multi-cellular organisms that must capture and ingest their food. This kingdom includes animals of all sizes and shapes, including sponges, jellyfish, worms, snails, crayfish, and innumerable insects. The vertebrates are just one small but well-known sub-group, which includes fishes, amphibians, reptiles, birds, and mammals.

# II   FOOD CHAINS, FOOD WEBS, AND ECOLOGICAL PYRAMIDS

The basis of all food chains are the "producer" organisms, such as plants and some bacteria and protists. Producers convert raw materials (such as chemical nutrients in soil or water) and energy (from the sun) into food, which can be eaten by "consumer" organisms. A species that eats only plant material is called a herbivore, or first-order consumer. Something that eats a herbivore is a carnivore, or second-order consumer. (Omnivores eat both vegetation and animals.) After that, the next consumer would be a tertiary or third-order consumer, and so on. Each of these different steps up the food chain is called a "trophic level."

Decomposers (including fungi and bacteria) are organisms that get their nutrients from dead and decaying organisms. In so doing, they help to recycle the earth's nutrients, such as carbon, nitrogen, hydrogen, phosphorus, and calcium. These nutrients are constantly reused on earth, but we need energy from the sun to sustain life. Hence, the common ecological expression, "nutrients cycle, but energy flows."

Food chains are a simple method to show "who eats whom" in a specific habitat. In reality, though, most creatures eat many things (or get eaten by many things), so food webs are more realistic representations of what really happens. In most ecosystems, there are innumerable interconnections, so the food web is extremely complicated.

There are usually far more producer organisms present in an ecosystem than primary consumers, and even fewer secondary or higher order consumers. If you could weigh the mass of organisms or the amount of stored energy at each trophic level, you would notice a similar trend. These patterns are sometimes called ecological pyramids, since the values for the number of individuals (or biomass or energy) all become smaller as you get higher up the food chain.

**Sample Food Chain**

**Fourth-Order Consumers**
*(e.g., loon)*

↑

**Third-Order Consumers**
*(e.g., small fishes)*

↑

**Second-Order Consumers**
*(e.g., aquatic insects)*

↑

**First-Order Consumers**
*(e.g., zooplankton)*

↑

**Producers**
*(e.g., phytoplankton & plants)*

# III COMMON AQUATIC INVERTEBRATES OF FLOWING AND STILL WATERS

**Freshwater sponges**
Porifera (*still water and slow currents*)

**Flatworms**
Platyhelminthes (*dark bottom sediments, various habitats*)

**Roundworms**
Nematoda (*bottom sediments, decaying vegetation, various habitats*)

**Segmented worms and leeches**
Annelida (*still water and slow currents*)

**Water mites**
Arachnida (*still water*)

## MOLLUSKS/MOLLUSCA

**Snails and limpets**
Gastropoda (*still and flowing waters*)

**Clams and mussels**
Bivalvia (*still and flowing waters*)

## CRUSTACEANS/CRUSTACEA

**Crayfish**
Decapoda (*various habitats*)

**Water fleas**
Copepods and Daphnia (*still water*)

**Side-swimmers, sandhoppers, sea lice**
Amphipods and Isopods (*still water and under rocks in fast water*)

> **NOTE:** This is intended as a very generalized guide. There are many different species within the groups, each varying somewhat in shape, size, and preferred habitat.

## INSECTS/ INSECTA

*(many aquatic forms are larval or "nymph" stages in the lifecycle)*

**Mayflies**
Ephemeroptera (*still and flowing water*)

**Dragonflies**
Odonata, Anisoptera (*still water and slow currents*)

**Damselflies**
Odonata, Zygoptera (*still water and slow currents*)

**Stoneflies**
Plecoptera (*flowing water only*)

**True bugs, including water striders, walking sticks, giant water bugs**
Hemiptera (*still water and slow currents*)

**Dobsonflies, alderflies, and fishflies**
Megaloptera (*still water*)

**Beetles, including whirligig and predaceous diving beetles**
Coleoptera (*still water and slow currents*)

**Caddisflies ("shadflies")**
Trichoptera (*still and flowing water*)

**Blackflies**
Diptera, Simuliidae (*flowing water*)

**Mosquitoes**
Diptera, Culicidae (*still water*)

**Craneflies**
Diptera, Tipulidae (*still water*)

**Midges (including "bloodworm," a red larval midge)**
Diptera, Chironomidae (*still, sometimes stagnant, water*)

# IV   GLOSSARY

**ACID PRECIPITATION:** Form of atmospheric pollution that harms both terrestrial and aquatic ecosystems; primary sources are sulfur dioxide and nitrogen oxides from the combustion of fossil fuels and metal smelting.

**ALIEN SPECIES:** Exotic species not native to a particular region; many are invasive and out-compete native wildlife and plants.

**AQUATIC:** Relating to water or water ecosystems.

**AQUIFER:** Underground water reservoir; found within layers of permeable rocks, sand, and gravel.

**BIOACCUMULATION:** Increase in concentration of pollutants in an individual organism over its lifetime.

**BIOCHEMICAL OXYGEN DEMAND (BOD):** Amount of oxygen required to decompose organic matter in a specific volume of water.

**BIOCONCENTRATION:** Increase in toxic pollutants in the tissues of living things. Bioaccumulation usually refers to increasing concentration over one individual's lifetime; biomagnification (or bioamplification) refers to increasing concentrations in organisms higher up the food chain.

**BIODIVERSITY:** A measure of the number and variety of different species found in an ecosystem; also, the genetic variation within species and the diversity of global ecosystems. The loss of the earth's biodiversity is a major environmental concern.

**BIOLOGICAL INDICATOR:** Bioindicator (or bioindex) species reflect the environmental health of their ecosystem. For example, the lack of certain species from an ecosystem could indicate pollution.

**BIOMAGNIFICATION:** Increase in the concentration of pollutants in organisms higher up the food chain. Also called bioamplification.

**BIOME:** Natural region (or bioregion) with characteristic plant and animal communities (e.g., boreal forest or temperate rainforest).

**BIOREGION:** Natural region with characteristic plant and animal communities.

**BOG:** Peat-dominated wetland which contains stagnant and acidic water.

**BOTTOM LANDS:** Low-lying and often relatively flat land in river valleys, subject to periodic flooding.

**COMBINED SEWER:** Sewer that contains sewage, household waste water, and rain runoff from streets and driveways.

**COMMUNITY:** Represents all living and interacting organisms in an ecosystem, including plants, animals, fungi, bacteria, and protists.

**CONIFEROUS:** Cone-bearing trees, such as pines, spruces, firs; usually evergreen, with some exceptions (such as tamarack and larch).

**COMBINED SEWER OVERFLOW (CSO):** Untreated sanitary and storm water that overflows from the sewer or treatment plant and enters lakes or rivers untreated.

**DECIDUOUS:** Broad-leafed trees, such as maples and oaks, which shed their leaves each autumn.

**DECOMPOSERS:** Organisms such as fungi and bacteria which obtain their energy by decomposing dead organic matter. Essential for the cycling of nutrients in ecosystems.

**ECOLOGICAL RESTORATION:** Any effort that aims to restore or expand natural ecological communities, thereby improving habitats and possibly also helping in pollution control.

**EDGE HABITAT:** Area where two different ecological communities meet.

**EMERGENT PLANTS:** Plants rooted in shallow water but have their leaves and some of the stem above the water's surface.

**EPILIMNION:** Warm upper layer of lakes; in summer, usually contains higher oxygen concentration than hypolimnion.

**ESTUARY:** Region where a river flows into the sea, often in a partially enclosed bay.

**EUTROPHICATION:** Process of nutrient enrichment in an aquatic habitat, which can lead to oxygen depletion. Can occur as a result of natural or human causes.

**EXOTIC SPECIES:** Alien species not native to a particular region; many are invasive and out-compete native wildlife and plants.

**FEN:** Peat-containing wetland, dominated by grasses, sedges, and some moss species, but not sphagnum; usually alkaline with at least a slow flow of fresh water.

**FLOOD PLAIN:** Land in river bottom lands, subject to annual or occasional flooding.

**FOSSIL FUEL:** Oil, gas, coal, and related fuels that are formed from fossilized organisms.

**GREAT DIVIDE:** The height of land on the North American continent that separates waters flowing into the Pacific Ocean from those flowing into the Atlantic or Arctic Oceans or the Gulf of Mexico. Located in the Rocky Mountains throughout most of the United States and Canada.

**GREENHOUSE EFFECT:** A natural process in which greenhouse gases, such as carbon dioxide and methane, trap heat from the sun in the earth's atmosphere. Pollution is increasing the amount of greenhouse gases and affecting global climate.

**GROUNDWATER:** Reserve of water occurring underground, within sand, gravel, or soil; may flow slowly through the ground.

**HEADWATERS:** Upstream region within a watershed; higher elevation source for rivers.

**HYPOLIMNION:** Cool bottom layer of lakes; in summer may become depleted of dissolved oxygen.

**INTERIOR HABITAT:** Relatively uniform habitat, distant from the edge.

**ISLAND BIOGEOGRAPHY:** Ecological theory used to predict the number of species on islands, based on size and distance from the mainland; also used to model community size and structure in isolated patches of terrestrial habitat.

**MARSH:** Shallow wetlands which usually remain covered with water year-round; not stagnant.

**METALIMNION:** Middle layer of a lake, where temperature drops rapidly from warm surface layer to cold deeper layer.

**NITROGEN FIXATION:** Process by which atmospheric nitrogen gas ($N_2$) is converted by microorganisms into a form usable by plants.

**OLIGOTROPHIC:** Refers to lakes that have relatively low amounts of nutrients and life; common in deep, cold northern lakes.

**OUTFLOW:** Downstream end of a river where it empties into the ocean or a lake.

**OVERTURN:** Mixing of surface and deep waters in lakes in spring and fall; occurs in seasonally stratified lakes when temperatures become uniform throughout; replenishes dissolved oxygen in deep waters and brings nutrients to surface waters.

**OZONE ($O_3$):** Protective layer in the upper atmosphere, shielding the earth from harmful ultraviolet (UV) radiation. At ground level, it is a form of human-caused atmospheric pollution or smog.

**PHOTOSYNTHESIS:** Process used by plants and some other producer organisms to make food from sunlight, carbon dioxide, minerals, and water; oxygen is a by-product.

**PHYTOPLANKTON:** Small, often microscopic, free-floating producer organisms in aquatic ecosystems.

**POOL:** Deeper, slower-flowing section of a river or stream; alternates with shallower, riffle habitats.

**RESPIRATION:** Cellular respiration is the process used by organisms to obtain energy from foods; carbon dioxide is a by-product.

**RIFFLE:** Shallower, fast-flowing section of a river or stream; alternates with deeper pool habitats.

**RIPARIAN:** Habitat found alongside rivers and streams.

**SANITARY SEWER:** Sewer system for sewage and other household waste water.

**STORM SEWER:** Sewer system for rainwater runoff from streets, driveways, rooftops, etc.

**SWAMP:** Flooded wetland characterized by standing trees.

**TERRESTRIAL:** Relating to land or land ecosystems.

**THERMOCLINE:** Zone in lakes separating the epilimnion and hypolimnion, characterized by rapid decrease in temperature with increasing depth.

**TRIBUTARY:** Stream or river that feeds into larger systems.

**TROPHIC LEVEL:** Refers to an organism's position within the food chain (e.g., producer, first-order consumer).

**WATERSHED:** Region that drains into a specific body of water such as a river or lake (e.g., Mississippi River watershed). Also called a drainage basin. Sometimes refers to the high-ground dividing line between distinct drainage basins.

**WATER TABLE:** Upper limit of an underground water reserve.

**WETLAND:** Habitat that is saturated with water all or at least part of the time, often because of relatively flat topography; includes swamps, bogs, marshes, fens.

**ZOOPLANKTON:** Small, often microscopic, free-floating consumer organisms in aquatic ecosystems.

# V  SELECTED BIBLIOGRAPHY

Andrews, William A. *A Guide to the Study of Freshwater Ecology*, Scarborough, ON: Prentice-Hall Canada, 1972.

———. *Investigating Aquatic Ecosystems*. Scarborough, ON: Prentice-Hall Canada, 1987.

Audubon Society. *The Audubon Society Field Guide to North American Fishes, Whales and Dolphins*. New York: Alfred A. Knopf, 1983.

Banfield, A.W. Frank. *The Mammals of Canada*. Toronto: University of Toronto Press, 1974.

Barnes, Robert D. *Invertebrate Zoology*. Fifth Edition. New York: CBS College Publishing, 1987.

Boraiko, Allen A. "Hazardous Waste: Stirring Up Trouble." *National Geographic* 167 (1985): 318–351.

Borror, Donald J., and Richard E. White. *A Field Guide to the Insects of America North of Mexico*. Boston: Houghton Mifflin Company, 1970.

Burt, William H., and Richard P. Grossenheider. *A Field Guide to the Mammals*. Third Edition. Boston: Houghton Mifflin Company, 1976.

Campbell, Neil A. *Biology*. Third Edition. Redwood City, CA: Benjamin/Cummings Publishing, 1993.

Dobson, Clive. *Feeding Wild Birds in Winter*. Toronto: Firefly Books, 1981.

Environment Canada. *What We Can Do for Our Environment*. Ottawa: Ministry of Supply and Services, Government of Canada, 1990.

Fitzharris, Tim, and John Livingston. *Canada: A Natural History*. Toronto: Penguin Books Canada, 1988.

Freedman, Bill. *Environmental Ecology: The Ecological Effects of Pollution, Disturbance, and other Stresses*. Second Edition. San Diego: Academic Press, 1995.

Gerster, Georg. *Amber Waves of Grain: America's Farmlands from Above*. New York: Harper Weldon Owen, 1990.

Godfrey, W. Earl. *The Birds of Canada*. Revised Edition. Ottawa: National Museums of Canada, 1986.

Graves, William, ed. *Water: The Power, Promise, and Turmoil of North America's Fresh Water*. Special Edition of the National Geographic Society (November 1993), Washington, D.C.

Hawke, David J. *Wetlands*. Toronto: Stoddart, 1994.

Hough Woodland Naylor Dance Limited and Gore and Storrie Limited. *Restoring Natural Habitats: A Manual for Habitat Restoration in the Greater Toronto Bioregion*. Toronto: Waterfront Regeneration Trust, 1995.

Keating, Michael, and the Canadian Global Change Program. *Canada and the State of the Planet: The Social, Economic, and Environmental Trends That Are Shaping Our Lives*. Toronto: Oxford University Press, 1997.

Labaree, Jonathan M. *How Greenways Work: A Handbook on Ecology*. Ipswich, MA: U.S. National Park Service and QLF/Atlantic Center for the Environment, 1992.

Lincoln, Roger J., and Geoffrey A. Boxshall. *The Cambridge Illustrated Dictionary of Natural History*. Cambridge, UK: Cambridge University Press, 1987.

Miller, G. Tyler. *Environmental Science: Working with the Earth*. Sixth Edition. Belmont, CA: Wadsworth Publishing Company, 1997.

McCafferty, Patrick W. *Aquatic Entomology: The Fisherman's and Ecologists' Illustrated Guide to Insects and Their Relatives*. Boston: Science Books International, 1981.

Mitchell, John G. "Our Disappearing Wetlands." *National Geographic* 182 (1992): 3–45.

Niering, William A. *The Audubon Society Nature Guides: Wetlands*. New York: Alfred A. Knopf, 1985.

Noel, Lynn E., ed. *Voyages: Canada's Heritage Rivers. Tenth Anniversary Volume*. St. John's, NF: Breakwater, 1995.

Peterson, Roger T. *A Field Guide to the Birds: A Completely New Guide to All the Birds of Eastern and Central North America*. Boston: Houghton Mifflin Company, 1980.

Peterson, Roger T., and Margaret McKenny. *A Field Guide to Wildflowers of Northeastern and North-central North America*. Boston: Houghton Mifflin Company, 1968.

Petrides, George A. *A Field Guide to Eastern Trees*. Boston: Houghton Mifflin Company, 1988.

Raffan, James, ed. *Wild Waters: Canoeing Canada's Wilderness Rivers*. Toronto: Key Porter Books, 1986.

Ricklefs, Robert E. *The Economy of Nature*. Second Edition. New York: Chiron Press, 1983.

Schaefer, Vincent J., and John A. Day. *A Field Guide to the Atmosphere*. Boston: Houghton Mifflin Company, 1981.

Scott, William B., and Mildred G. Scott. *Atlantic Fishes of Canada*. Toronto: University of Toronto Press, 1988.

Smith, Robert L. *Elements of Ecology*. Third Edition. New York: HarperCollins, 1992.

———. *Ecology and Field Biology*. Fifth Edition. New York: HarperCollins College, 1996.

Whitton, B.A., ed. *River Ecology*. Berkeley, CA: University of California Press, 1975.

Williams, D. Dudley, and Blair W. Feltmate. *Aquatic Insects*. New York: C.A.B. International North America, 1992.

# INDEX